DIRT RICH LIVING

FROM CORPORATE CUBICLES TO UDDER CHAOS

DIRT RICH LIVING

FROM CORPORATE CUBICLES TO UDDER CHAOS

MORGAN GOLD

A POST HILL PRESS BOOK
ISBN: 979-8-89565-319-7
ISBN (eBook): 979-8-89565-320-3

Dirt Rich Living:
From Corporate Cubicles to Udder Chaos

Cover design by Conroy Accord

All events, people, locations, and conversations depicted in this book are true and accurate to the best of my memory. But let's face it, the human memory isn't infallible, especially after decades of questionable life choices. Some names and identifying details have been changed or adjusted out of respect for privacy, friendship, sworn blood feuds, nondisclosure agreements, hidden treasure tontines, and a general desire to remain welcome at the annual Peacham Fourth of July Tractor Parade.

Post Hill Press
New York • Nashville
posthillpress.com

Published in the United States of America
1 2 3 4 5 6 7 8 9 10

CONTENTS

INTRODUCTION

LIVING DIRTY

The mudroom door groans open. The air is colder than it should be. Forty-two degrees in July still feels wrong. The farm is wrapped in a thick quilt of fog. The edges of barns and trees blur into white nothingness. The world feels unfinished, like I woke up before the day had time to load.

Before I can take two steps, Pablo Barn Cat appears on the porch railing, watching me like some old, disapproving landlord. He narrows his eyes, and I swear he shakes his head. Pablo Escobarn thinks he runs this place. Maybe he does. He was our farm's first animal, after all.

A blur of fur and claws scrambles up my jeans. Ginny Barn Cat, the aggressively friendly one, scales my T-shirt and takes up her usual lookout post on my shoulder. I let out a half-laugh, half-wince as her tiny claws burrow into my flesh. She nuzzles my neck, purring louder than a diesel engine. I am not a cat person.

There's commotion over by the farm gate. Toby Dog and Abby Dog are pacing, their tails cutting slow arcs through the mist. They look like night-shift security guards waiting to deliver their security briefing. Toby Dog lets out a low, deliberate huff. This sound makes me imagine he's clearing his throat before launching into a full report. *Coyotes sniffed the perimeter. No major incursions. The barn cats remain insufferable.*

Most people see hundred-pound, fluffy dogs and think *pets*. Not these two. Toby Dog and Abby Dog have jobs: Keep the farm safe from anything that fancies a duck dinner. And they take it seriously.

I pat their heads before I hop on my e-bike and ride up the hill. If there's one way to make morning chores a little more fun, it's zipping around on two wheels, mist curling around me like I'm gliding through some lost valley. Abby Dog lopes beside me, ears perked, ready for whatever comes next. My work commute used to involve traffic or subways. This is much better.

Hearing our approach, my fold of Scottish Highland cattle emerges from the fog in slow motion. Massive prehistoric creatures with shaggy coats and horns like broadswords. There's darn near twenty of them these days. They look tense, restless, and very heavy metal. Macho Man Randy Savage, the bull, bellows in irritation. *Fresh grass! Cream of the crop!* Deep bellows echo across the pasture in response.

I unroll a fresh strand of electric fence wire and step back, watching them rush to the fresh grass. Grazing, trampling, fertilizing in a perfectly efficient cycle. Right on cue, the cleanup crew arrives behind them in yesterday's paddock. The chickens are already pecking through the cow patties, hunting for fly larvae. Micro-raptors on macro-manure patrol, digging for treasures.

And then there are the ducks and geese.

I'm always late. They're always dramatic. They appear in the orchard, forming a single-file parade as they waddle toward breakfast. Their little feet slap the ground in unison, their quacks and honks of protest carrying through the fog. A sixty-foot-long, single-file horde of determined waterfowl warriors follow me to the place where I'll dump their food. If they ever figure out how to march in formation, we're all doomed.

By the time I finish up morning chores, the summer sun has burned off the last of the fog. The farm finally feels awake. I lean against the fence for a minute, taking it all in. The dogs, the cats, the cattle, the birds, the pigs, the goats, the bees, the trees...the land itself. A rhythm I didn't create but somehow stepped into.

But if you had told me ten years ago that this would be my life? That I'd be up at dawn, shifting cattle and dodging goose attacks instead of sitting in a windowless office reviewing marketing budgets? That a guy who grew up in suburban Connecticut and went to art school would be a farmer?

I would have thought you were crazy. How the heck did I end up here?

People love a good "starting over" story.

Fresh starts, clean slates, reinvention. It's the stuff of self-help books and TED Talks, the promise tucked into New Year's resolutions and midlife crises. Society eats it up. The idea that you can just *decide* one day to become a new person, walk away from your old life, and step into something better.

The problem? It's a lie. Because nobody actually knows how to begin.

Beginnings are never clean. They're tangled, clumsy, and full of wrong turns. No cinematic moments where you buy land and ride off into the sunset here.

No, the real beginning starts in the middle of a mess. It sneaks up on you. One day, you're fantasizing about a new life. The next, you're knee-deep in pig poops, wondering why you did this to yourself.

I was certain that quitting my corporate job would mean quitting my flaws, anxieties, and self. Tabula rasa!

Turns out, moving to a farm also doesn't magically change who you are as a person. I still wake up thinking about meetings and metrics. I still write to-do lists as if I were planning a product launch. I still feel the irrational urge to check my email, even though my inbox now consists mostly of feed-store coupons and spam from a guy trying to sell me alpacas.

(Side note: Don't get alpacas. I'm pretty sure alpacas are a multilevel marketing scam.)

People think clarity comes before action. They need to know *precisely* what they're doing before they begin. That's a lie. You don't gain clarity *before* you start. You gain it *by* starting.

That's the brutal truth of beginning: You must *be bad* before getting good. And that's a hard pill to swallow when you've spent your entire professional life optimizing for competence and fearing failure.

In my old life, I was used to *knowing things*. I had expertise. A skill set. A résumé full of bullet points that proved I was capable. But out here, none of that matters. The cows don't care that I can give a kick-ass pitch presentation. The chickens have no interest in my analytical abilities. The soil isn't interested in hearing my strong opinions about kerning.

There's something uniquely humiliating about being a grown adult and having to Google things like *how to fix a prolapsed duck penis*. Or realizing, after two hours of chasing escaped piglets, that you left the gate open in the first place.

That's the tax you pay for starting over.

You don't get to skip the part where you're bad at it. You don't get to avoid the moments that make you feel foolish. And most people, when they realize that, walk away before they ever truly begin.

Because the hardest part of beginning isn't the work. It's surviving *yourself*. Surviving the doubt, the second-guessing, the deep-seated fear that maybe you just weren't meant to do this.

The ones who make it, the ones who actually start over and stick with it, aren't necessarily the ones with the most money, or the best plan, or the perfect circumstances. They're just the ones who can push through that uncertainty long enough to get to the other side.

And if there's one thing I've learned, it's this: The only way out is through the mud and the muck.

Back when I lived in the city, dirt was something you avoided. It represented disorder, discomfort, something to be cleaned away and hidden beneath layers of concrete and carefully manicured grass. Dirt symbolized failure. A messy stain to erase, evidence that you'd lost control.

But here, dirt has taught me something completely different. It's become my greatest teacher, showing me that life isn't about avoiding the messy parts but leaning into them. Dirt isn't something to overcome. It's something you cultivate, turning uncertainty into growth, mistakes into wisdom.

And maybe farming isn't your thing. Maybe the idea of waking up at dawn to chase runaway ducks sounds like a nightmare rather than a dream. That's fine. This isn't a book about how to farm.

This is a book about what happens when you decide to mess up your life and do something wildly outside your comfort zone. Whether you're thinking about starting a business, going back to school, moving across the country, or simply pressing reset on a part of your life that no longer fits, the fundamental truths remain the same. You're going to get dirty.

Because at its core, beginning is never really about *what* you're doing. It's about *how* you handle the messy parts. It's about stepping forward even when you don't feel qualified, embracing the awkward, painful, and sometimes laughably bad early attempts, and learning to trust yourself in the process. No matter the goal, the hardest part is always the same: convincing yourself to start before you feel ready.

This book isn't a neatly organized farming manual filled with numbered checklists and foolproof formulas. Instead, what you'll find here is a chaotic collection of stories from someone who jumped without a parachute and lived to tell the tale. Maybe you'll find it helpful, whatever path you might be on. Even if you never set foot on a farm.

And now with all that blah-blah-blah said, let's begin.

PART ONE

ESCAPING THE RAT RACE

CHAPTER 1

WHEN YOUR DREAM JOB BECOMES A NIGHTMARE

As the mediocre Hartford skyline receded in my rearview, I felt my anxiety grow. Hartford was once the insurance capital of the world, but these days, most people seemed to drive around there without it.

Even with the uneasy feeling in my gut, it was nice to be back in my old hometown for the day. I had moved to New York City four years ago to take a new, bigger job for a lot more money. And even though Hartford was only about one hundred miles away, it felt like a different planet. I rarely go back to Hartford to see family and old friends. But on this day, I was in Hartford on business. It really felt more like a mission.

Connie, my boss, was seated beside me in the passenger seat of my rented Hyundai Elantra. An energetic woman in her sixties with tidy blonde hair, her presence was as commanding as it had always been. Connie had been my boss for several years at two different companies. I owed much of my rapid career rise as a financial services marketing executive to her.

As I drove, she rambled on and on about a recent customer-retention marketing program or something like that.

"It was fine," Connie said dismissively, waving a manicured hand. "But honestly, just more blah-blah-click-click blather if you ask me."

"Connie," I interrupted the words, feeling like marbles in my throat. "Yeah, so, I need to, uh…I've been, uh, thinking about some stuff."

She turned, her gaze settling on me with a calm expectancy, the kind that had always made room for my thoughts, no matter how trivial.

"What's on your mind, my dear?"

I took a deep breath.

"I need to resign," I said.

The words hung between us, heavy with the weight of betrayal. When I first started working for Connie, I was an unremarkable art-school kid who had recently taken a job as the "video guy" at a local insurance company to pay off a pile of student loan debt. A few short years later, I was now one of the youngest senior executives at the Fortune 100 investment company where I worked, overseeing a large part of the marketing organization.

Connie's reaction was a silent, wide-eyed stare. "You're joking," she finally managed, the disbelief in her voice a sharp pang in my chest. She quickly turned her attention to her ever-present and bejeweled iPhone.

That disbelief, that inability to conceive my departure as reality, wounded me more than I anticipated. Connie had been more than a mentor. She was like my work mom.

"No, Connie, I'm not joking," I confessed. "I've been offered a job as the head of marketing at an investment company in DC. It's been one of the hardest decisions of my life, but I think it's finally time for me to step out on my own."

"Why?" she finally asked.

I took a breath. Then another. The kind of pause you take before telling someone something they already know but don't want to hear.

And then I said it. The whole thing. The ambitions. The growth. The inevitability of it all. How I needed to carve out a future that wasn't just a natural extension of the past. How sometimes, staying felt like shrinking.

She listened, arms crossed, her expression giving nothing away. Then, slowly, her posture loosened. A shift so small I might've missed it if I hadn't been waiting for it.

Understanding. Maybe even acceptance.

"I never wanted to disappoint you," I admitted, the guilt a tight knot in my throat.

"I won't pretend this isn't a shock. But I've always known you were destined for more. If this is the step you need to take, then I support you."

Her words were a balm, soothing the raw edges of my guilt and offering a glimmer of redemption amid my turmoil. We reached her house, and as she got out of the car, she leaned back in, her eyes locking onto mine with an intensity that spoke volumes.

"Make me proud," she said softly.

After dropping Connie off at her home, the drive back to the city was a quiet one. As Hartford faded into the distance, a chapter closed with every mile that rolled beneath the wheels. I caught the Sox game on WTIC until I got too far south and had to switch to the FAN and endure the Mets.

Hartford is often relegated to a third-tier city. Too many big city problems to be quaint, too small a city to be relevant. The kind of place people drive through on their way to somewhere else. Which, fine. I get it. But when you live there, you learn to love its character and stubbornness. It had been my home, my career, my life. And for a long time, it was where I thought I'd stay.

But it was also where I met Allison.

We had met on an online dating site back when online dating was considered wicked sketchy. This was pre-app era. No swiping. No algorithmic soulmate matching. Just the digital equivalent of writing a personal ad on a diner napkin. With her willowy frame and a cascade of brown hair that always seemed to catch the light just right, Allison moved through life with an effortless, captivating grace. Her laughter brightened even ordinary moments. But it was her intellect that truly set her apart. Sharp as a tack, she could unravel the most complex ideas with a simple, elegant flair that left me in awe. Standing beside her always felt like winning the lottery.

Allison and I dated for about a year before it came time for her to leave Hartford. She had been working as a nurse in the emergency room of the local children's hospital, but she dreamed of working in international public health...you know...like USAID type stuff. She had been accepted into a grad-school program in New Orleans and was getting ready to move away. I knew I didn't want to be without her.

So I proposed.

She said yes, but on one condition.

"Anywhere but here."

I had once dreamed of spending so much time living in Hartford that I would one day become mayor. Allison hated the place. That was it. If I wanted to be with her, I had to get us out.

We agreed that she would be willing to follow me to wherever my career would take me, city-wise, as long as it was from a list of five cities: New York; Boston; Washington, DC; Atlanta; and Chicago. Five cities where she felt reasonably confident that she could find a job. While Allison was away at school, it would be my job to find a job outside Hartford and in one of those five cities.

Two weeks later, the universe played along. Connie, my boss at the insurance company where I had worked in Hartford, announced she was leaving to become the chief marketing officer at a massive investment company in New York. As a result, I was living in Brooklyn before summer was up.

Taking this new job at the investment firm in DC felt like stepping onto a moving train: fast, ambitious, going somewhere. The company had that rare, intoxicating mix of underdog hunger and legitimate financial muscle. This outfit shouldn't have been able to compete with the Wall Street titans yet somehow kept winning. I liked that. It made me believe we could build something special.

For the first few weeks, I rode the high. The office had that thrumming energy, the kind that makes you check your phone at night, not out of obligation, but because you actually *want* to see what's happening. Meetings felt dynamic. People moved with purpose. Ideas seemed to matter.

And then, reality started sharpening at the edges.

The CEO was always there. Always. And involved in everything.

He wasn't a *boss* in the traditional sense, more like a perpetual presence, a force field of scrutiny. He wasn't just involved; he was *embedded.* His office door was always open, not for accessibility, but as a watchtower. His eyes flicked constantly from his screen to the glass walls of his office, monitoring the movement on the floor.

At first, I thought it was just a hands-on leadership style. But then I noticed the patterns.

My email drafts? Reviewed.

Presentation font choices? Debated.

The color scheme for a digital ad?

"Let's run it in grayscale first, just to see."

The man, had strong, strong opinions, about commas.

One afternoon, I was sitting in my office when I felt him before I saw him. A shadow stretching across my desk.

"Quick thing," he said, stepping inside without knocking. He leaned on the edge of my desk, arms crossed, wearing the kind of expression that made you brace yourself. "That campaign you presented this morning. Solid work. But…"

Ah. There was always a "but."

"…I was thinking about the tagline. 'Invest Boldly'? It's good. *Really* good." He paused, letting the compliment sit just long enough to make me want to believe it. Then: "But what if we went with 'Bold Investments for a Bold Future' instead?"

I blinked.

"That's…three times as many words," I said carefully.

He laughed, as if I had made a joke.

"Right, right," he said, nodding. "But don't you think it adds *gravitas*?"

Gravitas. A word he used at least five times a day.

I could already see where this was going. A meeting would be called. Everyone would discuss the tagline. Opinions would be voiced, notes taken, options debated. The final decision? His, always his.

And so, over time, the initial thrill dulled.

I had taken this job thinking I'd be steering the ship. I would be the boss.

I would craft bold campaigns, shaping the company's voice. Instead, I was starting to feel like a human whiteboard, scribbling down ideas only for them to be erased and rewritten by the little man in the glass office down the hall. I ran the company's marketing in title alone.

I was just another set of hired hands executing someone else's vision. I was paid well, but the money felt like the only reward.

I tried. I really did.

There were things to like about DC. I threw myself into them, hoping to shake the gnawing sense that something wasn't right.

Allison and I made the most of our weekends. Sometimes that meant escape. Driving to Shenandoah to hike, heading to Annapolis for seafood, searching for pockets of quiet where life exhaled a little. Other times, it meant slowing down so much we practically dissolved into the couch. Lazy Saturday mornings, coffee in hand, the sun streaming through the windows, no agenda beyond existing in the same space. We had more time together than we'd had in years, and I reminded myself that this was the trade-off. Work less, live more. This was supposed to be the dream.

And yet.

I needed something. A creative outlet. A thing that was mine. And because of this need, all the half-baked creative projects I had locked away since leaving college started to come out of storage. I started writing again. Mediocre screenplays that were perfect in my head and absolute trash on paper. I bought fancy pens, convinced myself I'd finally crack that graphic-novel idea that had been gathering dust in the back of my brain. But the same cycle played out, every time. Excitement.

Ambition. Then? Distraction. Frustration. Abandonment. The unfinished projects stacked up like unopened bills, each one a reminder of something I used to be better at.

Work was lighter here. I wasn't drowning in it the way I had been in New York. Fewer hours, less travel, a schedule that should have felt like freedom. Instead, I felt unmoored. No longer drowning, but adrift, weightless, and not in the fun astronaut way, more in the lost-at-sea, no-land-in-sight way. My life lacked purpose.

And beneath it all, something settled in. A weight. A quiet, creeping kind of sadness, grain by grain, so gradual that I didn't even notice until, one day, I realized I was carrying more than I could bear. I was more depressed than I had ever been, and I had no idea how to shake it.

Or how much longer I could keep pretending I was OK.

The office was quiet, the soft hum of computers blending with the distant city sounds filtering through my window. I was in my office, reviewing a marketing plan I intended to present to my boss, when the door creaked open. The CEO entered, clutching a draft ad my team had prepared for an upcoming campaign. The ad was covered in sticky notes.

"Morgan, I've gone through your draft," said the CEO. "There are quite a few areas that need tightening up."

He dumped the papers on my desk. The printing deadline was tomorrow morning. I nodded, masking my frustration.

"I thought we were ready for a final review," I said in my best attempt at a customer service voice. "What's wrong?"

The CEO pulled up a chair and flipped open his notebook. I reflexively rubbed my forehead to calm my nerves.

"Let's start with the font. Futura? Far too pedestrian. And the margins? Make them narrower. We need to project sharpness, the cutting edge. That's communicated by proper margins."

I clenched my jaw, forcing back my true reaction. Sharpness? From margins? Really? Are people auditing mutual fund ads with rulers now?

"Details, Morgan," he said with a finger wag. "Details make or break perceptions. Now, about the budget, why is there so much for digital? We need more in print media. I read *Barron's* every Saturday morning."

"But the data shows our audience has largely shifted online. It's cost effective," I whined.

The CEO scribbled something down, dismissive. "I want to see alternatives. And more tagline options. This one isn't dynamic enough. Needs to be punchy. And memorable. More memorable. We pay too much money for these ads for them to just be forgotten."

I leaned back, the fatigue setting in. "I thought we had agreed that this was how we would go for this campaign."

"Yes, but it doesn't wow me anymore. I'll know it when I see it," he declared, standing to gather his notes. "Make these changes, and we'll review again. Remember, I need to approve everything before any external communication."

I didn't move.

The CEO had left, but his presence lingered like a rancid fart. The faint hum of his voice still rattling in my skull, the phantom weight of his hand pressing down on the campaign I had spent weeks refining. My fingers hovered over the keyboard, poised but paralyzed. Every tweak, every note on my work felt like another notch in a tally of compromises, a running inventory of how thoroughly my identity had become entangled in someone else's narrow vision because they're the one signing the checks, including my own paycheck. My time on this planet is finite and short. Is this really how I want to spend it?

Each of his "suggestions" had landed like a chisel against stone, chipping away at my sense of agency, my creativity, my point in even being here. This role, the one I had chased, strategized for, *uprooted my life for,* felt less like a leadership position and more like a carefully padded cell. The walls just happened to be made of PowerPoint decks and budget meetings.

The door clicked shut behind him, and I exhaled, long and slow.

And for the first time—no, not the first time, let's not lie to ourselves here—I asked the question I had been shoving into the farthest corner of my mind:

Is the prestige, the title, the absurd amount of money I am being paid worth the slow erosion of my sense of self?

I knew the answer. And the answer sucked.

The professional trophies of career success I'd chased had all the lasting joy of winning a claw-machine plushie: briefly thrilling, ultimately pointless.

At first, I had arrived at the company brimming with ideas. Big, loud, market-shifting ideas. Only to watch them get systematically downsized, defanged, diluted into something safe. Budget realities, sure. But mostly? The CEO's gravitational pull toward the *status quo* and his own ideas.

And so the job had become an endurance test. How much could I push before the leash snapped back? Going to prison for bludgeoning a man with a conference-room chair wasn't the future I had hoped for when I took this job.

And how much longer was I willing to pretend that didn't matter?

CHAPTER 2

FARMERS' MARKETS ARE FOR SUCKERS

It's a sunny Sunday morning in June, and the Dupont Circle Farmers' Market is putting on its weekly performance. A full spectacle of broccoli, baguettes, and babies in meltdown mode. The air has that humid DC clinginess, but a merciful breeze nudges through, teasing the canopies of vendor stalls that are a riot of colors, smells, and the occasional overenthusiastic dog bark. The humble farmers seem happy to have left their farms and homes for the day to go to the big city to work their booths and peddle their goods

Ahead, the scent of fresh basil mingles with the earthy aroma of locally grown carrots and beets, piled high in rustic wooden crates. The colors are a feast for the eyes. Deep purples, vibrant oranges, and the lush green of leafy vegetables. A farmer, with hands gnarled like the roots of the plants he tends, offers samples of crispy radishes, and you crunch one. The bite bites back with a peppery sharpness that makes you cough politely and nod as if to say, *Yes, that's exactly the experience I hoped for.*

The farmer is no big-bellied-overalls-sporting-corn-fed stereotype. Clad in cuffed jeans and a faded Guided by Voices tee, his arms are inked with obnoxiously minimalist tattoos. A constellation of stars, a sprig of rosemary, the words "Try Harder" in a sterile sans-serif font. A pair of wire-rimmed glasses balances on his nose, and his hair, tousled just so, suggests hours spent mastering the perfect "effortless" look.

Somewhere nearby, a woman with a septum piercing and a ukulele sings a song that's probably original but feels familiar. People tap their feet and pretend they aren't checking their phones to see if she's on Spotify. It's wholesome, but performative. Think Coachella, but for people who compost.

"This entire place is a scam," I declare to my wife.

Allison keeps walking and admiring the various market products. She ignores my cranky words with a measured sense of practice.

We approach a stall where a guy who looks like a rejected casting choice for a Vice documentary is arranging cuts of beef in a cooler like they're fine jewelry. The sign boasts words like "grass-fed," "organic," and "free-range," each one carrying a price tag that would make Peter Luger blush.

"Twelve dollars a pound for ground beef?" I say, keeping my voice low but not low enough. "That's not a price. That's a ransom note."

Allison gives me a look, the one that says, *Do you hear yourself?* Then she picks up a vacuum-sealed package of ribeye and turns it over in her hand.

"See anything in small print about the cow getting weekly massages?" I ask. "They use fancy words and sales pitches to make you pay more, but the product they're selling is no different than stuff from Safeway."

The farmer hears me. Of course he hears me. He smiles, patient in the way only someone secure in their moral high ground can be. "This beef is from cattle raised humanely, fed a grass-only diet. You can taste the difference."

"That's what they say about almond milk," I reply. "But that stuff just tastes like regret and xanthan gum."

Allison doesn't flinch. "Ignore him," she says to the farmer. "He has unresolved issues with food elitism. What he really wants is a Quarter Pounder."

"I don't want a Quarter Pounder," I counter, even though if I was being honest, I did want a Quarter Pounder. I probably wanted three Quarter Pounders. "I just want people to admit there's no functional difference between a twelve-dollar-a-pound burger patty and the kind you get for two bucks at the grocery store. Both cows mooed the same. It's all just marketing, and *I know marketing*."

It started, as most changes do, with routine. Sunday mornings became market mornings, not because I had a sudden change of heart about artisanal tomatoes or the moral superiority of grass-fed beef, but because Allison loved going, and I loved Allison.

At first, my role was passive. I was the mule, carrying the NPR tote bags and trailing behind her as she inspected crates of vegetables and frozen packages of meat like a jeweler inspecting uncut gems. But over time, a strange thing happened: I started paying attention, too. It was hard not to. All the sights and sounds and tastes.

One Sunday, I picked up a bundle of carrots, their greens still attached and a little unruly. I turned them over in my hands. They smelled like something. Not the sanitized nothingness of the plastic-wrapped kind in the grocery store, but real dirt, the earth they'd grown in.

"Don't get attached," Allison teased, her tone light, but her thread-plucked eyebrow arched.

"I'm just looking," I said.

We had begun cooking more at home. Not because I'd suddenly found religion in the fresh produce aisle, but because when you spend more money on your food ingredients, you get inspired to do more ambitious things with those ingredients. Boneless skinless chicken breasts grilled to leathery toughness on an old George Foreman are all fine and good for cheap protein orbs purchased in Styrofoam packs at the Aldi. But if you spend $28.93 on a whole chicken, you want to do it justice. And when you do your food justice, you taste the difference.

And while cooking was definitely step one, my true shift from skepticism to curiosity began with Mike, a sheep farmer we met at the market. He stood out, not because of his booth, which was modest, with just a few coolers of frozen lamb neatly displayed, but because of his easy demeanor. In his mid-fifties, he looked more like someone who should be tweaking CAD drawings than tending livestock.

"You were an engineer?" I asked, half incredulous, half validated.

"Yeah," Mike said, adjusting his cap. "Nineteen years designing systems for manufacturing plants. But one day, I realized I was building efficiency for things I didn't care about. So, I quit. Bought this farm. Learned how to manage animals in a way that heals the land."

He explained the basics of regenerative grazing, his voice calm but passionate. "The idea is to mimic how wild herbivores used to roam. Constantly moving, never overgrazing one spot. It's all about working with biology instead of against it."

Allison, intrigued as always, peppered him with questions about his methods, and before we left, Mike handed us a flyer. "Open farm day is next weekend. Come by if you're curious."

The following Saturday, we drove out to Mike's farm in Rappahannock County. Unlike the idyllic pastoral scenes you see on labels, the place wasn't particularly picturesque. It was practical. And functional. Rolling emerald pastures divided by electric fencing, a barn carefully organized with tools and feed, and a herd of sheep that moved in tight clusters, grazing deliberately as if they'd been trained to do so.

Mike greeted us with a wave and began the tour. "Regenerative grazing isn't just about raising sheep. It's about building soil," he said, crouching to pull up a clump of grass. The roots were long and thick. "With conventional systems, like the feedlots and confined animal feeding operations, or CAFOs, you're essentially extracting from the land. Animals are packed in tight, fed grains they aren't built to digest, and the waste becomes pollution instead of fertilizer."

He stood, brushing dirt from his hands. "Here, we're cycling nutrients back into the land. The sheep eat the grass, their manure fertilizes the soil, and the pasture improves yearly. Healthier soil holds water better."

I'd heard these ideas before. Buzzwords like "carbon sequestration" and "sustainability." But seeing it in action was different. This wasn't marketing fluff. It was science, ecology, and practicality working together.

Mike pointed to his infrastructure: lightweight electric fencing that allowed him to move the sheep daily, and a portable water system that followed the flock. "Mob grazing," he explained, "keeps the animals moving, which prevents overgrazing and lets the plants recover. The soil gets aerated naturally, and we don't need synthetic fertilizers or pesticides."

He mentioned Allan Savory, a Zimbabwean livestock farmer whose TED Talk on reversing desertification had garnered millions of views and sparked a global movement, and Joel Salatin, the outspoken Virginian farmer who turned his family's worn-out land into a model of profitable regenerative agriculture. "It's not alchemy," Mike said. "It's just understanding how biology works and mimicking what had worked in nature for thousands of years."

In the wild, large herds of herbivores (bison on the North American plains, wildebeest on the African savannas, and caribou in Arctic tundras) moved in dense clusters to protect themselves from predators. This instinctive behavior concentrated their grazing in small areas for short periods before the herd moved on. The land they left behind thrived: Their hooves pressed seeds into the soil, their dung and urine enriched it with organic matter, and the vegetation had time to regrow before the animals returned to start the cycle again.

Today, farmers mimic natural herd movements using electric fencing. By restricting the animals to graze intensely but briefly in each area, farmers prevent overgrazing while allowing plants to recover.

At the end of the tour, Mike invited us to watch him move the sheep. With nothing more than a bit of fencing and a whistle, he corralled the flock to a new patch of pasture. The animals' eager grazing felt almost ceremonial, like they understood their role in the system.

Driving home, I couldn't stop thinking about the sheer elegance of it all. It felt logical and beautiful. The kind of logic that made me wonder why we'd ever chosen feedlots and monoculture over systems that worked with nature instead of fighting it.

That night, over lamb chops we'd picked up from Mike's store, I felt something I hadn't expected: pride. Not just in the meal, but in knowing where it came from, how it was raised, and why it mattered.

"You're quiet," Allison said, breaking the silence.

"Just thinking," I replied, slicing into my plate. "It's amazing, isn't it? How it all works. How simple it is."

"So you're saying the beef at the grocery store isn't the same as Mike's lamb?"

"Correct," I said, savoring the bite. "Lamb and cattle are different animals."

Mike's farm didn't inspire curiosity in me. It sparked an obsession. The dopamine delivery of diving down a rabbit hole and learning about something new has always proven irresistible to me. Regenerative agriculture was no exception. Back home, I found myself crashing into this world with the fervor of someone who'd just discovered a conspiracy, only this one wasn't built on paranoia. It was built on hope.

I started with Allan Savory's TED Talk, the one Mike had mentioned. In just over twenty minutes, Savory explained how mimicking the natural movement of grazing animals could reverse desertification and heal degraded lands. It was a revelation, watching the before-and-after images of barren landscapes transformed into lush, green pastures. Despite some icky white-savior vibes and a serious lack of scientific data backing some of his claims, Savory's methods have inspired countless farmers, myself included. His words carried a weight that lingered: We can fix this. We just have to work with nature instead of against it.

From there, I stumbled into the orbit of Joel Salatin, the self-proclaimed "lunatic farmer" whose Polyface Farm in Virginia had become a blueprint for regenerative farming. His writing was equal parts sermon and practical guide, with titles like *Pastured Poultry Profit$* (not a typo!) and *You Can Farm*. Salatin's voice was brash, unapologetic, and occasionally *wicked problematic*. But beneath the rhetoric was a deep respect for the land and the creatures that lived on it. And for me, in my earliest days of being farm curious, it was very inspiring.

What struck me most was the accessibility of it all. This wasn't about sprawling industrial operations or cutting-edge technology. It was about observing patterns, respecting biology, and using simple tools to create a system that thrived. The methods were humble—electric fencing, portable water troughs, rotational grazing schedules—but the results were profound.

Even the science (not Savory's pseudo-science, but the real, tried-and-true methods) pulled me in. I had always dismissed phrases like "carbon sequestration" as buzzwords, but now they felt urgent and real. I learned how healthy soil acts like a sponge, pulling carbon dioxide from the atmosphere and storing it underground. Regenerative grazing wasn't just good for farmers. It was a potential solution to some of the biggest environmental challenges we face.

I devoured books, podcasts, and YouTube videos. I learned about compost tea and cover crops, symbiotic relationships between fungi and plant roots, and the ways regenerative methods could outpace conventional farming in yield, profitability, and resilience.

Meanwhile, my day job began to feel more alien. Sitting in meetings, scrolling through spreadsheets, and debating deadlines...I couldn't shake the feeling that I was pouring energy into something that didn't matter. The seemingly simple rhythms of farm life called to me. Every decision Mike made, every fence he moved, was part of a system that created something tangible and meaningful.

For weeks, I binged videos and articles about gardening, learning terms like "raised beds" and "composting" with the same enthusiasm I used to reserve for fantasy football. Allison, ever the realist, watched my growing obsession with mild amusement.

"You know this means ripping out the lawn, right?" she said one morning as I scrolled through online seed catalogs.

"I know," I replied, full of conviction. "It's wasted space."

By the next weekend, I was knee-deep in the dirt, shovel in hand, carving out sections of our postage-stamp-sized front yard to make room for garden beds. At the time, Allison and I lived in a beautiful, modernly restored row house in DC's Capitol Hill neighborhood. The block was lined with old row houses, leafy trees, and the occasional flower garden. At the time, there wasn't a vegetable patch to be found.

The neighbors watched from their porches, their expressions mixed with curiosity and disbelief. Elmo, my next-door neighbor, eyed my activities with confusion.

"You're really tearing up that nice grass?" he asked.

"It's not nice grass," I replied, wiping sweat from my brow. "It's mostly weeds."

"Well, I hope you know what you're doing."

I didn't, but that was the point.

The first weeks were a blur of trial and error. I built makeshift raised beds, filled them with soil and compost, and planted rows of tomatoes, kale, and squash. I added herbs along the edges, including basil, thyme, parsley, partly for cooking and partly because they made me feel like a farmer.

When the seedlings I had started in the house during winter were transplanted to the beds once things got warm, I felt an absurd sense of victory. By midsummer, the garden was in full swing, with tomato plants sprawling out of control and cherry tomatoes bursting in clusters. I'd wake early to check on the plants before work, and every time I saw something new, such as a flower, a ripe fruit, or a visiting bee, I felt a quiet thrill.

The garden wasn't just about food. It was about possibility, about proving to myself that I could create something tangible and real. It felt like a small act of defiance against the convenience-driven world I'd always accepted without question.

The first tomato we harvested was celebrated with the kind of ceremony usually reserved for birthdays. I grilled it with olive oil and salt, served it alongside a fresh tomato salad, and declared it the best meal I'd ever made. Allison charitably didn't argue.

The garden became my gateway drug. With each small success, I started dreaming bigger. Sketches of potential backyard expansions filled the margins of my notebooks. I toyed with the idea of adding a greenhouse, researching everything from materials to climate-control systems.

"You realize this isn't a farm, right?" Allison said one evening as I rattled off plans for crop rotation.

"Yet," I replied, grinning. "Not a farm *yet*."

But the idea of a farm was no longer a joke. I found myself scrolling through listings for small plots of land, imagining rows of crops and animals grazing in the distance. Mike's voice echoed in my mind: It's not magic; it's biology.

On weekends, Allison and I started taking drives to rural areas, just to see what was out there. We talked about the logistics of buying land, of starting small and scaling up. She humored me, as always, but I could tell she was starting to see the spark that kept me going.

One evening, as we sat in the backyard with dirt still under my nails from pulling weeds, I looked out over our little garden and imagined it scaled a hundredfold: rows of vegetables, a small flock of sheep, a life lived closer to the land.

"What if we had more space? What if we bought a farm?" I asked.

"A farm?"

"Yeah. Not to live on full time, but like a vacation place or somewhere we can retire early to."

"OK."

"Really?"

"Sure. Why not? We don't have kids. We have plenty of money saved. Gardening and hockey seem to be the only things making you happy these days. Let's do it!"

I was stunned. I didn't expect her to be so agreeable to the idea. I had been planning this conversation for weeks, expecting hard pushback.

"Is this crazy?" I asked, half to myself.

"Probably," Allison replied, her voice steady. "But it's a complicated and misunderstood kind of crazy. Like Frances Farmer."

CHAPTER 3

I HOPE TO BE EATEN BY A BEAR

I know how I want to die.

In the distant year of 2084, I'm heartbroken as my beloved Allison has peacefully passed away in her sleep. Her death marking a century well-lived. Her departure, peaceful and dignified, marks the end of an era and the beginning of my final adventure.

After taking a few days to put all our affairs in order, the last testament to our life together, I wake up one morning and make myself a sandwich. Nothing fancy, just something to munch on. Maybe roast beef, maybe ham, and definitely lots of mustard. The fancy brown kind with crunchy mustard bits.

With my brown bag packed, I head out to the familiar pasture of our farm. I do my 104-year-old-man shuffle through the fields that I've worked for decades. Maybe I'll have a cane or a walker. Or maybe, because it's the future, I have a mech suit like some sort of Robotech character who just happens to work in agriculture.

The world around me is hushed, save for the scratchy whisper of late-September leaves skittering across the ground and the occasional heckle of a blue jay. I head straight for the butternut tree, which was not just any butternut tree, but *the* butternut tree, a gnarled, time-worn monolith that has stood sentinel over this land long before people farmed this land.

Settling down at its base, I lean my wrinkly old-man skin against the rough bark. I can feel every nook and cranny press into my spine like an old friend who never learned about personal space. The air smells of damp earth. I close my eyes, let the weight of the tree anchor me. I don't know if trees remember things, but if they do, this one has seen it all.

The hours slip by, measured in the slow unwrapping of my sandwich, in the lazy arc of the sun as it climbs, hesitates at its peak, then starts its inevitable retreat. The shadows stretch long and thin, like cats in the late-day heat, curling and twisting around me.

And then, without fanfare—no dramatic crack of a branch, no eerie shift in the air—a black bear steps into the clearing. A massive black bear. The kind that looks like it walked straight out of some primordial dream. Its coat ripples, dark as ink, its eyes catching the light like twin flints sparking in the dusk. It sees me. Not just glances. *Sees.*

I rise to meet it, heart hammering in my ribs, but somehow, I'm grinning. "Well, old friend," I whisper, the words feeling like they've been waiting years to be spoken, "let's see what you've got."

And the bear, either humoring me or indulging in the grand theater of the wild, charges. The impact is thunderclap and avalanche, a primal, absurd communion of muscle and instinct. We move like we've done this before, like we've always done this. I feint left, dart right, the bear a blur of teeth and sinew, of power and play. It's a fight, yes, but also something more, an untamed, reckless waltz choreographed by whatever ancient force decides these things.

And for the first time in a long time, I feel alive.

In the end, the bear overpowers me, as it is meant to be. I am consumed and my body returns to the earth, part of the eternal cycle of life and decay. Scavengers will later clear the scene, leaving behind little more than a story. As they say, we're all just compost in training.

Whenever I tell strangers about this dream, they are often Werner Herzog–level horrified. But when I reflect on this imagined end, I see it as a story of triumph. If I can meet my end like this, at such an age, then I've indeed done something right in life.

From that imagined final day, we spool the tape backward, back to the present, where my life as a farmer isn't about endings but about everything that builds toward them. Farming, for me, isn't just about coaxing things from the dirt. It's about tending. To the land. To relationships. To the stubborn, necessary work of getting up and doing it all over again. It's about being so tangled up in the rhythms of nature that they stop feeling like something outside of me and start feeling like a reflection in the mirror. Seasons shift, new life is born, some go to slaughter, crops rise and fall, and somewhere in all that repetition, I see the shape of my own life, looping, spiraling, growing. *Ka is a wheel.*

Spring is the start. The moment it stirs, stretches, and shakes off its winter hangover. Everything wakes up hungry. The soil is thick with the ghosts of last year's crops, the decomposition of things that once thrived, now turned to fuel. Decay isn't the end of the story. It's the start. Seeds crack open, roots stretch, and suddenly, the world smells alive again.

Summer? Summer is chaos. The fields go wild, green, sprawling, and relentless. Plants climb over each other like kids on a jungle gym. The air hums with life, everything buzzing, breathing, growing. But even in all the lushness, you can feel something lurking ahead. The plants don't get to grow forever.

Then comes fall, a season that smells like endings. The fields go quiet, stripped bare, and the skeletons of harvest are left to wither and collapse. But even in death, they're doing the work, breaking down, dissolving, turning into something useful again. Nothing really vanishes. It just changes shape. The nutrients slip back into the dirt, setting the stage for whatever comes next.

And then winter. The great pause. Everything above ground looks dead, but underground, the earth is plotting. The soil is soaking up everything the last seasons left behind, metabolizing the past into something the future can use. It's quiet, still, patient. Because it knows what's coming.

It always does.

This perpetual cycle in agriculture of planting, growing, harvesting, and decaying mirrors the existential cycles we observe in our own lives: birth, growth, culmination, and eventual decline. Each phase is necessary, each end is a beginning, and nothing exists in isolation.

Understanding this helps me appreciate the transient yet eternal nature of life. It teaches us respect for the processes that sustain us and a reverence for the intricate dance of creation and destruction that bookends our existence.

But why choose a life that is undoubtedly hard, sometimes heartbreaking, and always dirty? Because in that grit, there's truth.

"I don't know what exactly we're waiting for," said Allison.

Allison and I were having one of those rare quiet dinners together at home. I had been sharing the usual frustrations about my day at the office. The stifling micromanagement, the endless tweaking of marketing campaigns that felt like alphabetizing the spice rack in a meth lab. That's when the conversation took an unexpected turn.

"Why do we keep doing this to ourselves?" she asked, setting down her fork, her voice cutting through my foggy discontent. "You're miserable. It's like you're becoming someone you're not."

I sighed, the weight of her words sinking in. "I know. It's just that… this was supposed to be the dream job, right?"

The glow from the overhead light spilled across the table. Her brow furrowed slightly.

"I don't know," she began, her voice tinged with fatigue. "Lately, even my work doesn't feel right anymore. It's like I'm just going through the motions, not really making the difference I thought I would."

I bobbed my head in understanding. "I've been feeling the same," I admitted. "We bought the farm as a getaway, but whenever we're up there, it feels more like home than here. What if we're living our lives waiting for a future we could have now?"

The previous year, after countless late nights scouring the internet, we had found it. The locals called it the "Old Shaw Farm" after the family that had owned the place for roughly a hundred years. It was 160-ish acres of rolling Vermont countryside, tucked into the Northeast Kingdom, where winter hangs on a little too long and the summers are the kind you dream about when you're stuck in city traffic. The town? Peacham. Population? Seven hundred, give or take. The kind of place where everyone knows your name and, more importantly, knows exactly what you've been up to.

The farm had been sitting on the market for six years, waiting. A massive farmhouse, the kind they don't build anymore, wrapped in a sagging but defiant porch that stretched around it like an old dog curling up for a nap. And then there was the barn. Three stories of ancient, timeworn dairy history, creaking but still standing, like it was daring us to bring it back to life.

We bought it for $350,000 in cash. Which is a lot of money, but not a lot of money for a farm of that size, even back in 2016. And yeah, I recognize how fortunate we were. Buying property outright isn't something everyone can do. Heck, these days just buying property in general doesn't seem accessible to a lot of folks. So did we luck out? Yup. Did we have an enormous financial advantage starting our farm without debt? Absolutely. But here's the kicker: Having a financial cushion doesn't actually make you fearless. It just swaps out one kind of anxiety for another.

The farm had once been a thriving dairy farm, passing through only four other owners since the 1830s, each leaving their mark, their stories, their ghosts. And now, somehow, improbably, it was ours. An investment, a vacation home, a potential future. As a joke, we had started calling it "Gold Shaw Farm." All it took was adding a *G* in front of *Old* on the sign out front.

Move to the farm now?

The idea hung between us, an unexpected seed magically taking root in mere seconds.

"We could make the farm more than just a retirement project," I suggested. "We've been renovating it, investing in it. Why wait?"

Allison paused, considering. The lines of stress softened around her eyes as she entertained the thought. "Are you saying we move there? Full-time?"

"Yes," I said, the conviction in my voice surprising even me. "We both know we're not happy here. The farm could be more than just an escape. We've been building it up as a place to live out our dreams eventually. Again, why wait?"

"But we can't just drop everything. We're not financially ready to retire, especially not in our thirties," she reasoned, ever practical.

"I know," I acknowledged. "I'm not talking about retiring. I'll find a new job. And you, you've been saying how you want to go back to school and become a nurse practitioner. Finding a medical practitioner job in the middle of nowhere is easy!"

Allison's eyes scanned our plans for the farm, her mind ticking through the logistics and possibilities. We didn't have a lot holding us back. We had always wanted to be childless by choice, so it's not like there were kids to worry about. We also had ample savings and investments.

Her voice trailed off, but then she lifted her eyes to meet mine, a spark of resolve flickering.

"But maybe this is what I need. A new challenge. A new way to make an impact."

"It's settled, then," I said, feeling a surge of relief and excitement. "We turn our backup plan into our main plan. We make a life where we don't need a vacation from."

That conversation changed everything. It wasn't just about leaving behind the noise, the burnout, the creeping sense that our lives had become a series of checkboxes. We were making a choice. Choosing to stop waiting, to stop pretending that "someday" was anything but an excuse. We weren't just running from something. We were now running toward it.

From that night on, everything felt different. The farm wasn't just a project anymore. It was *the* project, the place where our past selves and future selves finally met in the middle. It wasn't a someday thing. It was a right now thing. Some people have a calling. Me? I had a series of questionable decisions, impulsive detours, ADHD hyper-fixations, and clumsy mistakes that somehow added up to a farm. I didn't seek my purpose. I backed into it by accident. But it gave me a sense of purpose, nonetheless.

PART TWO

BAPTISM BY MANURE

CHAPTER 4

TREES GROW STRONG ROOTS

The apartment I was living in for my final few months in DC felt like a squat. A sad mattress pushed against one wall, a battered desk I found on the street crammed beside it, and a galley kitchen with a stove that aggressively buzzed when it turned on. Most evenings, I came home to silence, dropped my bag onto the chair, turned on a Caps/Sox game (season dependent), and peeled the lid off a takeout container, the plastic fork scraping against the plastic container as I picked at my dinner. Outside, the city droned on in the gray tones that painted the District of Columbia. But inside, it was just me, counting down the days until I could leave.

We had sold our rowhouse, packed up the life we had built, and whittled my possessions down to what could fit in this tiny space so I could keep my job in DC. At the same time, Allison moved to Vermont to return to school. When Allison and I had first met, she was working in international public health. Distributing AIDS and malaria medication in Nigeria, monitoring Ebola clinics in Liberia. Work that changed the world but burned her out. When we decided to move to Vermont, Allison decided to get her nurse practitioner's license to return to practicing medicine. Rural medicine had a dire need and offered virtually guaranteed employment.

Allison had a plan, a purpose, a new beginning. I had a folding calendar pinned to the wall, each square methodically marked with an X. Two weekends a month, I'd wake before dawn, toss a bag into the car, and drive six hours north, the city shrinking in my rearview mirror as the rolling hills of Vermont stretched open before me. Those weekends were everything. But each weekend inevitably ended, leaving me back in DC's gray liminality.

I tried to ignore the feeling of being unmoored, of floating between two places but belonging to neither. The job, the apartment, and the city were all temporary, just a means to an end. But the problem with always thinking of something as temporary is that it keeps you from being anywhere.

On those long, solitary nights, I read about permaculture. About the way trees anchor themselves, sending roots deep into the soil, shaping the land as much as the land shapes them. I thought about what it meant to truly belong somewhere. Not just to occupy space, but to invest in it, to build something that would outlast me. Trees weren't like me. They didn't move from place to place, always looking for the next opportunity. They stayed. They grew. *And maybe*, I thought, *it was time I learned how to do the same.*

I first came across Mark Shepard the way I seem to find most of my new obsessions, by tumbling down a rabbit hole I barely realized I'd stepped into. One day, I was a guy casually interested in regenerative agriculture. The next day, I was binge-listening obscure farming podcasts like they were the new true crime. Somewhere in that blur, Mark Shepard showed up. I don't remember the exact episode or host, but I remember when his ideas clicked into place.

Shepard isn't just a farmer. He's a kind of agricultural anarchist. He took worn-out farmland in Wisconsin and turned it into a thriving, self-sustaining ecosystem by doing something radical: listening to nature instead of fighting it. His book *Restoration Agriculture* wasn't just a lightbulb moment for me. It was a stadium's worth of floodlights all switching on at once. He wasn't just sustaining land. He was actively improving it year by year. His ideas weren't about eking out an existence but designing abundance.

Trees weren't just decorative. They were ecosystem anchors. Producing food, building soil, managing water, and sequestering carbon. Why bust your ass plowing fields every year when you could plant trees once and let them handle the heavy lifting?

And then I found out about chestnuts.

A hundred years ago, the American chestnut was the king of the North American forests. The nuts were starchy and nutritious, a perennial food crop. And then, like some kind of botanical horror story, it was gone. A fungus from imported trees hit in the early 1900s. And then, through a combination of the blight and human intervention to address the blight, nearly four billion chestnut trees were wiped out within a few decades.

And then I learned about people, Shepard included, who weren't just mourning the loss of the chestnut but actively bringing it back. Not as a conservation project. Not as a museum piece. As a legitimate, scalable, modern crop.

Because here's the thing: Chestnuts still work. Better than corn. Better than wheat. They don't need yearly replanting. They don't demand endless inputs. They hold the soil together instead of stripping it away. They produce food for a hundred years or more, pulling carbon from the air the whole time. They're one of the few genuinely perennial carbohydrate crops we have. A grain replacement that doesn't require sacrificing the land it grows on.

Imagine if, instead of endless acres of corn, we had chestnut orchards doing the same job without the erosion, chemical runoff, or the treadmill of replanting. That's the future I wanted to be a part of.

And that's how I became a tree guy.

But wanting it and making it happen were two different things. The dream of trees drove me forward, but the reality of getting there was overwhelming. Selling our DC rowhouse, job-hunting in a rural economy, and living five hundred miles away from my wife for months on end. The loneliness in DC was to be expected, but it was paired with an anxious undercurrent of uncertainty. Was I making a massive mistake? Could I really reshape my life around an idea?

I had no idea what I was doing. None. Zero. And Theodore, the consultant I had hired to help design the orchard, knew it. He was the type of guy you often meet in the permaculture space. You know, the kind of guy who thinks of every "weed" as a misunderstood friend. But also the kind of guy who collected bootleg Phish tapes and could MacGyver a bowl with only an apple and some aluminum foil.

"Wow, this spot here seems to really want to be a pond," Theodore would declare as we stumbled across the property. "You should really listen to it. Do you hear what it's saying?"

He rattled off observations about soil composition, wind exposure, and water runoff while I struggled to take it all in. I nodded, pretending I understood.

There were many possible choices. Where to put the trees, dig the swales, and leave things untouched. It was both exhilarating and paralyzing. Every decision felt enormous, not just because it would shape the farm but because it would be something I had to live with for decades. This wasn't like choosing an apartment or accepting a job offer. This was carving something into the land itself.

It didn't help that I've always found it difficult to make permanent choices. My wife and I could spend months, years even, debating which piece of art to hang on the wall, and now I was supposed to decide where an entire orchard would go? Where might a pond be dug? Where would I someday graze animals? It felt absurd. And yet, at the same time, the weight of those choices made the whole thing feel real. Terrifying but real.

Every book I read advised waiting a full calendar year before making significant changes, watching how the water moved and the seasons shaped the land. That was sensible, but patience is not my virtue.

I did the next best thing: I made a sketch, let it sit, changed it, let it sit again, and repeated this cycle until the anxiety became a background noise. I should have waited longer. If I could do it over, I'd have spent twelve months just watching the land breathe before I touched a thing. Instead, I planned and learned simultaneously, like trying to assemble IKEA furniture without looking at the instructions.

Eventually, though, I had to do something. The first significant step was cutting into the land itself. Theodore and I mapped out a series of swales and berms that needed to be dug to slow water runoff and help it soak into the ground rather than rushing away.

But digging required renting an excavator.

The *thunk* of metal against dirt echoed across the field. A second horseshoe spun in the air, caught the sunlight momentarily, and landed with a satisfying *clink* around the stake. It didn't matter. Only a few bored Holsteins watched us.

"Close counts. Just like in hand grenades," Matty Kempton called, shading his eyes with one hand as he grinned at me. A sturdy dairy farmer in his fifties, he had been paired with me during Peacham's annual Fourth of July horseshoe tournament.

The Fourth of July in Peacham, Vermont, was as American as a slice of apple pie, if apple pie came with the rumble of antique tractors and the scent of barbecue drifting through the air. I'm talking about the most quintessentially New England stuff imaginable. There's food and farmers' markets, games and activities, and the main attraction: one of Vermont's largest tractor parades. Hundreds of tractors wind through the streets, their drivers waving enthusiastically at crowds of well-wishers, celebrating America's birthday with pride and good cheer.

Matty stepped forward, his stance easy but practiced. He was broad-shouldered, sunburnt, with hands that had spent a lifetime working the land. His parents, George and Pat, had moved to Peacham in the early 1960s, and by that point, it had grown to the largest dairy farm in town. I watched as he lifted the horseshoe, gave it a lazy swing, then let it fly. It spun end over end.*Klink*. A ringer.

I let out a low whistle. "Alright, alright. Maybe it's not just luck."

He smirked. "Told ya."

Matty and his wife, Dawn, now ran the farm with the help of their adult sons, Will and Dylan. As Matty and I tossed horseshoes, our conversation turned to my farm and future plans.

"So," Matty said, tilting his head toward me, "what's the plan with your farm?"

I hesitated. "Still figuring it out. A lot to do."

Matty nodded. "Thought about buying that farm once," he admitted. "Didn't quite fit our needs, but it's good land." He took a sip from his cup, then added, "If you ever need a hand…the boys and I'd be glad to help."

It was a simple offer, spoken plainly, but it carried weight. This was the kind of community where neighbors helped neighbors.

"I'd appreciate that," I said.

And that's how, later that summer, Will Kempton arrived on the farm with his massive excavator. He climbed out, nodded in my direction, and gestured toward the field. "So where we diggin'?"

I unfolded my hand-drawn map, the lines crisscrossing the sloped pasture, marking where each trench would go. Swales, shallow ditches dug along the land's natural contour, would catch and slow down rainwater, letting it seep into the soil instead of rushing off in torrents, carrying nutrients away. The berms, mounds of earth piled on the downhill side, would create a natural barrier, helping retain that moisture while also providing the perfect place to plant trees.

Of course, knowing that in theory and digging nearly two miles of trenches were two different things.

Will climbed into the excavator like stepping onto an old bicycle. On the other hand, I had spent an embarrassing number of hours watching YouTube videos on operating one. My attempts had been jerky and imprecise, the bucket swinging wildly when I meant to make a small adjustment. Will made it all look easy, dragging the bucket through the earth in smooth, clean lines.

As he worked, I followed behind with a measuring stick, ensuring we stayed on contour. That was the key; if the swales weren't level, the water would pool in one spot instead of spreading evenly.

After about an hour, Will cut the engine and leaned out. "So, how do you know this is gonna work?"

"I mean, it works everywhere else," I hesitated.

He smirked. "Yeah, but have you ever done it before?"

"Nope."

He laughed, shaking his head. "Well, guess we'll find out."

By the end of the week, the once-smooth hillside was marked with curved trenches, each one following the natural flow of the land. Looking at them, I felt something shift in my gut, like I had done something real.

It still didn't look like much. Just a series of scars in the dirt. But soon, they'd be full of water. And later that fall, full of trees. Hopefully.

The minivan smelled like a lumberjack's gym bag. Sweaty, woodsy, and vaguely threatening. In the rearview mirror, a tangle of branches swayed with every bump in the road, their bare limbs stretching toward the windows as if the trees themselves were peering out, curious about the world they were about to join.

I had picked up the rental that morning in DC, signed the paperwork as casually as if I were just another guy shuttling kids to a soccer tournament, then immediately driven six hours west to western New York to a small permaculture nursery run by a guy called Akiva Silver.

Akiva wasn't just a tree grower; he was the kind of person who spoke about forests the way other people talked about old friends. I had read about his nursery, Twisted Tree Farm, obsessively, watched every video, and devoured his *blog posts* like they held the answers to everything I was looking for.

I first visited Akiva's farm earlier that spring to attend a workshop he taught about growing trees. During that first trip, camping out in Akiva's orchard, I learned that tree geeks are a special kind of person.

"You're going to plant your forest?" he asked, brushing soil off his hands.

I nodded. "Trying to."

He grinned like he had heard that one before. "You're gonna love these chestnuts."

Over the next hour, we dug out the trees I had pre-ordered: chestnuts, hazelnuts, apples, pears, plums, elderberries, and black locusts. He moved through the rows with the kind of ease that comes from knowing a place deeply, pulling up each tree, shaking the dirt free, and inspecting the roots before handing them off to me.

As I packed the trees into the back of the van, layering them carefully so the roots wouldn't dry out, I couldn't shake the ridiculousness of what I was doing. I had turned a rental minivan into a rolling forest. At a stoplight on the way back, I glanced in the rearview mirror and saw branches stretching toward the windows, a tangle of limbs swaying gently with each turn. It felt less like I was transporting trees and more like I had invited them along for the ride.

Somewhere on the highway, the doubt crept back in as the sky darkened and the reality of what I had taken on settled in. I had hundreds of trees in the back of this van. Hundreds. And in less than forty-eight hours, I was supposed to get them all in the ground.

By myself.

Well, not entirely. A few weeks earlier, in what might have been a moment of great foresight or complete desperation, I had sent out an email. "Planting weekend," I had called it. A lighthearted invitation: *Come to Vermont. Dig some holes. Drink some beer. Enjoy the fresh air.* The replies had trickled in. Some were enthusiastic, some vague, and some just a single thumbs-up emoji, which could have meant anything.

And that's when the real panic set in. Because this wasn't just about planting trees anymore, this was about people. And I have never been good at asking for help.

Help was unreliable. It was a door you opened without knowing if anyone would walk through. What if no one showed up? Or worse, what if they did show up, and I had no idea what to do with them? What if it rained, or the ground was frozen, or my plans were completely off-base? The thought of standing there with a van full of dying trees, a group of expectant faces waiting for instructions I couldn't provide, was horrifying. And even if everything went perfectly, accepting help meant owing favors. I hated owing people favors.

I gripped the steering wheel and forced myself to breathe.

One thing at a time.

That night, I pulled into the farm. I stepped out of the van, walked around to the back, and took a deep breath. Then I swung open the doors. The trees had made it. That was step one. I just had to figure out how to get them into the ground.

By Saturday morning, the house was full.

For the first time since buying the farm, every room in the farmhouse had people in it. My mom had claimed the guest room, my dad took the couch, and a childhood friend had squeezed an air mattress into the library. Friends from high school carpooled up from Connecticut, their trunks packed with sleeping bags and work gloves. Allison's college friends piled into the kitchen, helping her brew pot after pot of coffee.

And then there were the neighbors in Peacham. People I had barely spoken to. People who had nodded politely at the cafe but never lingered long enough to chat. Now they were showing up with shovels slung over their shoulders, work boots laced tight like this was some old-fashioned barn raising.

It was overwhelming.

I had spent so much time here alone, planning, pacing, overthinking, that suddenly being surrounded by so many people who had shown up just to help *me* it was almost too much. I kept waiting for someone to ask, "So, what exactly are we doing here?" in a way that implied they were questioning *me* rather than the logistics.

Instead, people just started working.

We set up an assembly line of sorts. My dad and a couple of friends tackled the digging, sinking their shovels into the damp spring soil, carving out space for each tree. My mom and aunt handled the root soaking, carefully untangling and hydrating the bare-root saplings before passing them along. Allison and her friends moved down the rows, placing the trees, making sure the roots spread naturally, tucking soil back in like they were tucking in a child for bed.

I hovered everywhere and nowhere, bouncing between groups, answering questions, but mostly just trying to make myself useful. The truth was, I felt like a fraud. I had read the books, I had watched the videos, I had drawn and redrawn the maps. But standing in the middle of all these people, I still felt like I was just making it up as I went along.

And then, at some point, I stopped overthinking and just started planting.

I dropped into the dirt beside my mom, who was pressing mulch around the base of a newly planted chestnut.

"Remember when I couldn't even keep a houseplant alive?" I muttered.

She smirked. "Oh, I remember." She patted the soil into place, then sat back on her heels, looking out over the orchard. Rows of young trees stretched in every direction, their plastic tubes swaying slightly in the wind. "But this? This is different."

And for the first time, I believed her.

By late afternoon, the last tree was in the ground. We stood there, sweaty and exhausted, staring at the work product. They still looked like sticks. Thin, fragile, almost comically small in the vastness of the field. But they were there.

That night, we built a bonfire. Nothing honors new trees better than burning dead ones. We passed around beers, plates of food, and stories. The mixture of friends and family from various parts of our lives came together to celebrate, leaving both Allison and me feeling a sense of gratitude that we felt like we had lost in DC.

"We did it," Allison whispered, nudging me gently. The firelight flickered in her eyes, mixing exhaustion and pride. "This place feels different now, doesn't it?"

For the first time, standing by that bonfire, I realized something crucial: Stepping out of my safe, predictable city life hadn't just brought me to a farm. It had brought me closer to myself and the people I loved. Maybe I could've found fulfillment another way, without uprooting everything and moving to the middle of nowhere, but I was certain I'd never have found it by clinging to my comfort zone. Comfort might keep you safe, but it rarely helps you grow. Sometimes, the only way to really know what you're made of is to throw yourself into something that makes absolutely no sense on paper and see what happens next.

CHAPTER 5

RELEASE THE QUACKEN

"I feel so alone here," she said.

The winter of 2017/2018 had been a hard one. Allison had moved up to our Vermont farm in the summer of 2017, brimming with the excitement of a life change. By January, she told me, snow had swallowed everything: the gravel drive, the stoic barn doors, even the idea of warmth. Nights dragged out endlessly, like the shadows of trees stretched thin over the snow, their branches clawing at the windows in the wind. Allison called often, her voice drifting like smoke through the line.

"It's so quiet here," Allison murmured one night, as though afraid to disturb the stillness by speaking. "And cold, too. I thought I'd like it, but…I feel invisible."

I wanted to tell her I felt invisible too, still boxed into my DC apartment where traffic murmured like an endless, indifferent tide. My days blurred together: meetings, emails, and sad evenings watching hockey games while I ate dinner alone.

The farm's potential haunted my daydreams. I pictured animals in the barn, their breath clouding the air in the mornings. Chickens, ducks, maybe sheep. I imagined hauling feed in buckets, breaking ice on the water troughs, hearing the comforting rhythm of life we'd build together. It felt so real in my head, but every time I opened my eyes, I was still six hundred miles away, stuck in limbo.

Allison was trying, even if the farmhouse seemed to resist her. She was back in school, working to become a nurse practitioner, a job that would make it easy for her to find work in Vermont's sparse rural healthcare system. To help with her loneliness, we even adopted Pablo Barn Cat, a cantankerous tabby cat who had been in need of a new home. He now seemed determined to claim the property as his own. But I could hear the strain in her voice when we talked.

"It's just hard," she confessed one evening, her words slow and deliberate, like a dropped stitch in a knitting pattern. "This place is supposed to be ours, but it's just me here."

"I know," I said, hating the distance that hollowed out my reply. "I'll be there as soon as I can."

That was all I had to give her. And time had all the patience in the world.

The months ticked by like frost creeping across a windowpane. Allison told me stories about Pablo's hunting conquests, finally fixing the sagging porch steps, and much more. But every tale felt like a reminder of what I was missing. She sent pictures of her snowshoeing trails, her cheeks flushed pink, the world behind her silent and glistening. She was trying to make it work. For both of us.

When I visited each month, the farm greeted me like an old quilt, heavy and familiar. There was woodsmoke in the air, and a stillness that wasn't empty but full of potential. I'd breathe it in, aching with the feeling that this was where I belonged. And each time I left, it hurt.

Then, in the Spring of 2018, it finally happened: The job offer I'd been chasing landed right in my lap. I had accepted a job running marketing for an insurance company in Vermont. I would make about half as much as I did in DC and it would be a two-hour commute to the office each day, but I would be able to finally live full-time on the farm.

I packed up my life in DC faster than a politician backtracking on a campaign promise, each box sealed tight with dreams of fresh Vermont mornings, open skies, and absolutely zero traffic jams. Unless you count cows blocking the road as a traffic jam.

The drive north felt like the first breath after being underwater too long. Mile by mile, the city faded behind me, replaced by quiet stretches on I-91 and green mountains that still had their snowy white caps.

I arrived in May. Just as Vermont's horrendous mud season had begun its slow retreat, unveiling bright green everywhere you looked. Allison sat on the porch, bundled in her thickest coat with Pablo napping on her lap. She smiled, quiet but radiant. I was home.

Moving meant I could finally get animals. We bought the farm in 2016. Most of 2016 and 2017 were dedicated to fixing up the house. The only agricultural activity I could partake in was planting the permaculture orchard. When you live nearly six hundred miles from your farm, you can manage trees, but having animals is the worst possible idea. Even though I didn't know much about farming back then, I knew that.

So, I waited.

Once I got the job in Vermont, it felt like my green light. These were going to be my first farm animals ever. Growing up in suburban Connecticut, the only pets we ever had were dogs, cats, and the occasional gerbil. I'd never had any experience with chickens, ducks, geese, goats, or other farm animals.

Allison and I sat in the farmhouse living room, the woodstove still going strong despite the fact that we were halfway through spring. We sipped tea from mismatched mugs. The moment was idyllic, so much so that I hesitated to break it with the thoughts swirling in my head. But I couldn't hold back any longer.

"So I think we're ready for animals," I declared.

Allison gave me a side glance, her expression deadpan.

"I've been waiting for this ever since we bought the place. Chickens, ducks. Maybe cattle or sheep one day. You know, actually having a farm with, well, farm animals."

She raised an eyebrow. "Uh-huh. Because we don't have enough to do already?"

"Exactly," I said, grinning. "But at least now we're here, not stuck dreaming from far away. It finally feels possible."

"Well, farm animals are famously low maintenance," she snarked.

"But I've really thought this through," I protested. "I'll start small. I swear! Conventional wisdom says poultry is the way to go."

"Let me guess," she said, tilting her head. "You're doing something unconventional."

"Ducks." I grinned.

She gave me a long, unimpressed look. "Of course."

"Hear me out!" I said quickly. "Everyone does chickens. Duck eggs, though? They're different. You can charge more for them, and there's less competition. I even had this idea of becoming the 'duck mogul of New England.' Think about it. Premium duck eggs in Boston, Portland, even the Connecticut suburbs. It's a lucrative market niche no one's tapped into."

Allison let out a snort of a laugh. "So, Ducks Unlimited but bougie?"

"Why not? It's better than being just another guy with chickens," I said, crossing my arms. "And let's be honest, duck eggs are richer, creamier, and just plain better."

At that point, I'm not entirely sure I had ever had a duck egg before. *(That's called foreshadowing!)*

"Sure," she said, her voice dripping with sarcasm. "What a universal truth. I can't tell you how many times I've been at the store, thinking, 'If only someone would bring me a high-end duck egg.'"

"You joke, but they're a premium product," I said. "This could actually work."

She smirked and shook her head. "Fine. If this ends in a reality TV series about you and your ducks, I want royalties."

By the spring of 2018, I was deep in research mode. Back then, there weren't as many resources on raising ducks as there are now, but I found a couple of YouTubers who really helped me. One was a guy named Matt, who ran a channel called *50 Ducks in a Hot Tub*. He had a massive duck operation, probably the biggest one I'd ever seen, and was constantly hatching ducks. Watching his videos gave me a sense of how a large-scale duck farm worked.

The other person who duckfluenced me was Jack Spirko. He had a farm in Texas and raised ducks for profit, selling their eggs. Jack made a series of videos on YouTube called *The Duck Chronicles*, where he documented the daily life of his ducks, particularly when raising new ducklings. Those videos were formative for me. They helped me understand the basics of duck care and how to think about my farm.

I was watching videos constantly, reading books, and falling further down the rabbit hole of duck research. YouTube was by far the most engaging and educational resource for me. Books were helpful, but they didn't capture the hands-on, practical knowledge I needed. It became clear that I was going to start my farming journey with ducks.

When you start looking into ducks, you realize there are so many breeds to choose from, each with its own characteristics. There are Muscovy ducks, which are incredibly hardy and great for free-ranging and hatching their own ducklings. There are runner ducks, which look like little bowling pins with legs and are excellent at catching insects and weeding orchards. Pekin ducks are large, white, and friendly but prone to overeating because they are usually raised for meat production. And then there's the Khaki Campbell. Most folks don't know this, but a Khaki Campbell duck lays about the same number of eggs per year as most egg-laying chicken breeds. But the duck is bigger and has proportionally more yolk. They're known for being excellent egg layers and seemed like the best choice for what I had in mind.

I ordered forty Khaki Campbell ducks. I planned to raise them as a "straight run," meaning the hatchery would send me about half males and half females. I planned to butcher the males for meat and keep the females as my laying flock. I figured twenty laying birds would be the minimum I needed to make selling duck eggs worthwhile. I placed my order online in April, scheduling delivery for early June, the earliest I felt it would be safe to raise ducklings on the farm. After placing the order, I spent the next few weeks in a flurry of preparation.

I converted an old milk-storage room in our three-story dairy barn into a brooder, an enclosed indoor area with heat lamps to serve as a momma duck for your little ducklings. I lined the corners with hardware cloth to keep out rats and weasels. I built a little enclosure to house the ducklings and researched everything I'd need, from heaters to waterers. Watching YouTube videos gave me tips and tricks on how to care for ducklings.

Around this time, I also started making YouTube videos. In a few short months, I would be flush with duck eggs, and I would need buyers. The people watching the videos could be potential customers. And documenting my experiences felt like a way to share this exciting new chapter with my friends and family, many of whom lived in cities far away and some of whom thought I had lost my mind.

"You've got a couple of boxes here, and they're loud as all get-out."

"I'm on my way," I replied as I hung up the phone.

The drive to the post office was short but felt endless, the anticipation pressing on me like an overstuffed bag. When I stepped through the door, the sound of frantic chirping hit me immediately, echoing through the small space.

Not many folks know this, but shipping baby poultry through the mail is a time-honored tradition in the United States, supported by the remarkable biology of newly hatched chicks, ducklings, goslings, and other poultry. These baby birds can survive for up to seventy-two hours without food or water, thanks to the yolk sac they absorb before hatching, which sustains them during transit.

"They're alive. For now," the postal worker said, his voice as dry as the air in the post office. He leaned on the counter, regarding me with the kind of detached curiosity reserved for patrons with packages that peep.

"Uh, thanks?" I replied.

He disappeared into the back room, leaving me to the muffled sound of frantic chirping. When he returned, he plunked two medium cardboard boxes on the counter. The boxes vibrated, chirping intensifying as if confirming their living cargo.

"They've been like this all morning. You sure you're ready for them?"

"Oh, absolutely," I said, lifting the boxes carefully. I wasn't sure he believed me, and to be honest, I wasn't sure I believed myself.

"Good luck," he muttered, watching me as I headed for the door.

In the car, the boxes pulsed with warmth and life beside me, their insistent chirps creating a soundtrack of chaos. The noise and commotion attracted the attention of our farm's new barn kitten who sat perched on the hood of Allison's car like a tiny gargoyle.

Lil Barn Cat hopped down, and immediately began sniffing around my feet. Her ears twitched at the sound of the boxes, and her eyes lit up with a gleam I didn't like.

"Don't even think about it," I warned, clutching the boxes tighter. She followed me anyway, tail swishing with barely contained excitement.

The brooder room was warm, the heat lamps casting a soft red glow over the carefully prepared bedding. I set the boxes on a table and opened the first one. A dozen pairs of tiny black eyes stared up at me, their bodies a chaotic mass of golden fluff.

"Welcome home," I murmured, lifting the first duckling out. Its body was soft and warm, and it squirmed in my hands, chirping indignantly. I dipped its beak in water, a step I'd read was vital, and placed it gently into the brooder. It stumbled for a moment before scampering off to investigate.

One by one, I repeated the process, moving methodically while Lil scratched at the door outside, her yowls occasionally cutting through the chorus of duckling chirps. "I hear you!" I called through the door. "You're not coming in, so give it up!"

Of course, she didn't.

When I finished, forty-two ducklings filled the brooder. Two more than I'd ordered. Hatcheries will often send a bonus bird or two "just in case." Luckily, all of them were alive, and only one looked a little sluggish. I stood back and watched them dart around, already making a mess of the water dish. But despite the chaos, watching those ducklings dash about brought unexpected joy, reminding me exactly why I'd chosen this life in the first place.

Lil Barn Cat's persistence, however, hadn't waned. As I stepped outside the brooder room, she tried to dart past me. "Absolutely not," I said, scooping her up mid-leap. She squirmed in my arms, clearly insulted.

Her yellow-green eyes locked on the brooder room door as if to say, *You can't keep me out forever.*

The first few weeks went smoothly. The ducklings were thriving. Almost too well, in fact. They started out as chirpy, fluff-covered blobs that you could cup in your hand. Now they had morphed into awkward teenagers with awkward feathers. They had also outgrown their brooder. It started clean and smelling of pine shavings. It was now poop-caked, smelly, and ready for an upgrade.

Enter: the chicken tractor. Or rather, the "duck tractor," because in my world, labels are more of a suggestion. Picture a mobile bird condo: an open floor plan, great foragers, and zero rent. The idea is as simple as it is genius, an open-bottomed cage you can scoot around your yard like some kind of backyard nomad. The birds get fresh grass and bugs on the daily, and your lawn gets a poop-powered spa treatment. Everybody wins, unless you're a blade of grass.

There are a variety of different types of chicken tractors that you can build, but the design I decided to construct was created by a guy named John Suscovich, a regenerative farmer from Connecticut, whose work had guided me during my early forays into farming. Back then, he was just a distant voice on a podcast and a face in online videos. Now, through a strange twist of time and shared passion, I can call him a friend. His chicken tractor design isn't flashy, but it's effective—sturdy enough to last, versatile enough for everything from meat birds to ailing ducks in need of a quiet recovery space.

Building the chicken tractor took effort, particularly given my extreme carpentry inexperience. But the process of living creatures depending on you to complete a project forced me to buckle down and do my best.

When I moved the ducklings out of the brooder and into their new setup, it felt like progress. They were feathering out, exploring their world, growing into themselves. Everything seemed fine.

And then, not long after, on the Fourth of July, I found my first dead bird.

"What happened?" I asked aloud, my voice trembling as I crouched near the still body of the duck.

The other ducks were running around doing duck things. But all of it felt distant, drowned out by the sight before me. Face down in the mud. There were no signs of a struggle, no predators, no wounds. Just death, sudden and inexplicable.

I knelt closer, hesitating, before brushing a feather from her neck.

"What happened to you?" I murmured.

The dead duck didn't answer. I swallowed hard and forced myself to stand, the weight of that single loss settling deep in my chest.

It shouldn't have felt like this. I had been warned, time and again, that farming wasn't for the faint of heart. Death was part of the deal. "Livestock means dead stock," they'd said. And I thought I was prepared. But nothing prepared me for the ache of finding her like that.

Over the next few days, others followed. I'd spot one stumbling, its movements sluggish, its head drooping unnaturally. Another would lie still in the shade, legs curled under its body as if simply resting. But they weren't resting.

"Not again," I muttered one afternoon, my hands trembling as I scooped up another lifeless body.

I was desperate for answers. At night, I sat hunched over my laptop at the kitchen table, scrolling through endless forums for backyard farmers and poultry keepers. The screen's glow was harsh in the darkness, the words on the pages both hopeful and maddeningly vague.

"Botulism," one thread suggested.

"Coccidiosis," another argued.

"Definitely duck herpes," a third chimed in.

I slammed the laptop shut in frustration. Each theory seemed plausible, but none offered certainty. The days blurred together as I tried everything I could think of—new feed, cleaner water, supplements. But the deaths kept coming.

One evening, I took matters into my own hands. With shaking fingers, I prepared for a crude necropsy. The body lay before me on an old towel, its feathers still soft and warm. I didn't know what I was looking for. Swollen organs? Signs of disease? Something obvious that would tell me how I had failed them?

But all I found was heartbreak.

By the end of the week, thirteen of my ducks were gone. Thirteen graves lined the edge of the woods, their tiny markers a painful reminder of how little I understood this life I had chosen.

Looking back, with the benefit of years of experience raising hundreds of birds, my theory as to what happened was that it was a niacin deficiency. Niacin. Good ol' $C_6H_5NO_2$. Ducklings need it in those critical early weeks of life, and my reliance on chicken feed had failed them.

Seven years later, I now use commercial duck feed that you can find at most feed stores. It's balanced for the niacin needs of ducks. And I haven't had similar problems since I made that change. But the weight of those first losses still lingers.

Farming is full of hard lessons learned. It can strip you bare, exposing the raw cost of inexperience and the inevitability of mistakes. It places you at the edge of life and death, forcing you to confront that fragile boundary more often than ever imagined. But farming also teaches resilience. The kind that roots itself deep, growing stronger with every hard lesson. If you're considering raising animals, understand this: Mistakes will happen. You'll lose more than you thought you could bear. The guilt will nag at you and make you feel like crap. But each loss carries its own lesson. And with every lesson, you grow stronger and get better.

But yeah, in July 2018, I lost thirteen ducks. It sucked.

When I think back to that first summer of farming, my rookie season, it feels like flipping through an old scrapbook. Every page was a snapshot: me hauling water at dawn, the sun painting the pasture gold, tiny fluffballs that would eventually grow to be quacking ducks.

Before the farm, I lived in a blur of deadlines and coffee-fueled commutes, where "slowing down" wasn't even a concept. But out there, with those little waddling poets of the pasture, I learned to see again. Their movements were frenzied yet somehow elegant, like a dance troupe that rehearses chaos, and they showed me the beauty in the mundane. Watching them became my daily meditation, a reminder that sometimes the most magical moments are tucked into the ordinary.

Not that it was all serene postcard scenes. Farming is messy. Literally. Ducks are poop factories. Cleaning their house felt less like a chore and more like a battle of wills. The routines of feeding barn cats, hauling water, and scooping poop became my gym membership.

And then there were the ducks themselves: comedians, philosophers, and revolutionaries rolled into one feathery package. Watching them waddle after slugs or stage their morning stampedes reminded me to laugh. Their camaraderie—always moving as one, like they were plotting a coup—made me admire their collective spirit. I joked about them forming the first Duck Communist Party, but their teamwork was a lesson in unity.

Getting my ducks to go to bed at night became an unexpected mission. I'd seen it done, with entire flocks waddling into their houses on command, so I knew it was possible. But knowing and doing are two very different things.

The first hurdle was my duck tractor itself. Ducks don't appreciate steps or thresholds; they see them as insurmountable challenges. So, with some advice from fellow farmers and a bit of creative engineering, I rigged up a funnel using garden netting and fence posts, which made for a janky-looking setup, but one that gave the ducks a clear path home.

The second hurdle was training. I had seen Jack Spirko train his ducks to go into their coop at night by yelling, "All ducks go to bed!" I wanted to do something similar, so every night, I'd call out, "All ducks go to bed!" to get my birds into the duck tractor for the evening.

At first, it was chaos. Some got it right away, while others wandered off like rebellious teens sneaking out past curfew. I used sticks to herd the stragglers. Slowly but surely, though, it started to click.

Night by night, they started to march inside the tractor with less and less fuss. It wasn't perfect, but it worked. And on a farm, that's what you celebrate.

As the season wound down, I realized I'd fallen in love, not just with the ducks, but with the rhythm of farm life. The grind that once felt daunting became a source of pride. People with ADHD brains like mine often chafe at the constant repetition of daily chores. But I found that chores became my favorite part of the day by slowing down and taking time to notice the little things, the wiggle of a tail feather or the glint of dew on a blade of grass. If you're thinking of farming, here's my advice: Embrace the grind. Show up. Marvel at the little things.

"Are you sure about this?" Geoff asked, holding a duck gingerly as if it might explode in his hands.

"Yes, I'm sure," I said, though my stomach was doing flips. "We've come this far. Besides, the ducks aren't going to butcher themselves."

Leslie adjusted her gloves and gave me a skeptical look. "You say that, but wouldn't that be something? Self-butchering ducks? Patent that idea."

"Right after I invent a duck that plucks itself," I muttered, picking up the knife.

It was fall, and my little ducklings that had arrived in the spring were now full-grown. Half were male, and the other half were female. Too many male ducks, or drakes, in a flock, creates chaos and suffering. Drakes are highly territorial and aggressive, particularly during mating season. Suppose the ratio of males to females is too high. In that case, the females are relentlessly pursued, over-mated, and often injured or stressed to the point of declining health. It disrupts the flock's balance, turning what should be a peaceful group into a scene of constant conflict. Keeping the proper ratio of one drake for every four to six females ensures harmony and the well-being of the entire flock. We needed to butcher approximately half of my ducks to balance my flock and put meat in the freezer.

My friends Geoff, Leslie, and Nick offered to lend me a hand on butchering day. Leslie's father, John, also joined us for the festivities. None of them had ever butchered birds before, but they were all very curious about the process. I was exceptionally inexperienced myself, having only learned through watching YouTube videos and helping out a friend butcher chickens earlier that summer.

The ducks waited quietly in the shed. I tried not to think too much about what we would do. This was my first time butchering anything, and I hadn't taken a class or had a mentor to guide me. I had YouTube videos and books and a whole lot of second-guessing.

"All right," I said, swallowing hard. "Let's do this."

The first bird was calm, looking up at me with the same unblinking expression it always had. Geoff placed it in the killing cone while I steadied the knife.

"It's important to be quick and clean," I said aloud, half to reassure myself.

"Yeah, please emphasize quick," Geoff said, shifting nervously.

With one swift motion, it was done.

"Not bad," Leslie said, breaking the silence. "You're, like, half a butcher already."

Nick grimaced. "More like ten percent. Let's not get cocky."

I sighed, a knot in my chest unraveling. "One down, a million to go."

The plucking, however, was a whole other beast.

"This is a nightmare," Geoff said, his gloves speckled with damp feathers. "Why is it like this?!"

"Because ducks are jerks," Leslie said, yanking at a particularly stubborn wing.

Plucking chickens is easy. Scald, toss, done. But ducks are waterproof. Their feathers are designed to repel everything, including hot water. We scalded. We scalded again. We turned up the heat. And still, the feathers clung stubbornly to the skin, mocking us.

By hour two, our morale was plummeting. We were resorting to brute force with needle-nose pliers, bare hands, and a lot of swearing. The feathers floated through the air, clinging to our clothes, hair, and souls.

"Who knew feathers could be so sticky?" Nick said, holding up a clump of down that refused to budge from his glove.

"Sticky is one word for it," I muttered, yanking another handful. "Evil might be another."

By the time we finished the last bird, my hands were raw, my back ached, and my brain was screaming at me to reconsider every life choice I'd ever made.

That evening, as the kitchen filled with the savory aroma of roasted duck, I sat down to reflect. The day had been brutal, but it had also been meaningful.

Here's the thing: Most people have no idea where their meat comes from. We're so far removed from the process of raising and killing animals that it's easy to think of meat as just another product. A boneless, skinless chicken breast wrapped in plastic. But it's not. It's a life. There's a moral cost to every piece of meat we eat. And standing in the middle of that process, feeling the weight of a living creature in your hands, the warmth of its body, the quiet stillness after it's done. It forces you to confront that cost in a way no supermarket ever could.

It's not easy. And it shouldn't be.

I wish more people could experience that moment. Not to turn them into butchers or farmers but to remind them that meat isn't just a commodity. It's a responsibility.

So, as I sat there, picking at the edges of my roasted duck, I made another promise: I'd learn from today. Next time, I'd be faster, better, more prepared.

And next time, I would figure out a better way to pluck a duck.

CHAPTER 6

A FARMER'S BEST FRIEND

One bright April morning in 2019, just as our harsh Vermont winter was starting to yield to the sloppy wet kiss of mud season, I stepped out toward the duck coop. This time of year is often more gray than golden, but that morning carried spring's upbeat vibes. Eager to start the day and shoot a new YouTube video, I approached the coop. I expected to hear the usual raucous chorus of quacks. But that morning, the coop was silent.

As I neared the coop, a sense of foreboding tightened its grip on me. My recently consumed breakfast was doing backflips in my belly. Peering through the coop window, the sight that met my eyes sent a chill down my spine. Samuel Puddleduck and Jemima Puddleduck, my cherished Pekin ducks, were splattered with crimson. The red made a stark contrast against their white feathers. I pressed my nose to the window's glass to look closer. Inside, the coop was a scene of carnage. It looked like someone pipe-bombed a pillow factory. Blood and feathers everywhere, a massacre of absurd proportions.

My heart pounded like a woodpecker trapped in a mailbox. I flung open the door, prepared to confront either a predator or some duck zombie apocalypse scenario. I honestly wasn't sure which was worse.

I stood there for a moment, feeling numb, as the morning sunlight spilled through the coop doorway and illuminated the aftermath in sharp detail. Loose feathers drifted aimlessly in the faint breeze, catching the sun's rays like misplaced snowflakes. A soft rustling broke my trance. A surviving duck stirred nervously, eyes wide with lingering terror. I swallowed hard, heart still racing, aware that my peaceful illusion of farm life had been shattered forever. The sanctuary I believed I'd built had just been reclaimed by the wild.

Amid the wreckage, something settled in—a mix of sorrow, fury, and the cold, unshakable certainty that this would never happen again. I had seen bad days on the farm, but nothing like this. Nothing that hit this hard. The air was thick with the metallic bite of blood, the ground littered with feathers that would never be preened again. Carnage. That's the only word for it.

The day became a blur of tending to the injured, cleaning up what couldn't be saved, and reinforcing the coop with a quiet, grinding determination. Every nail hammered into place echoed my internal mantra: This was my fault. Never again. Never again.

A strange calm had taken hold by the time the sun dipped low. Not peace, exactly. This wasn't just about fixing what was broken. This was about fortifying. About making damn sure that when the night came again, it wouldn't find us unprepared.

I poured all my efforts into fortifying the coop, sealing every potential entry point. No small crack or crevice was overlooked. Given the nature of the injuries and the chaos, my suspicions pointed toward a mustelid: perhaps a least weasel, or a mink. The damage didn't seem characteristic of larger predators like coyotes or bobcats, which would likely have left less destruction and struggled more to breach the coop.

Determined to protect my ducks, I set each trap with surgical precision. I positioned them around the duck house like a general fortifying a battlefield. This time, I would be ready. If that bastard came back, it wouldn't leave. As I lay in bed that night, a cautious relief settled over me. Not victory, not yet. But I knew I had done everything I could to secure my flock, and that let me sleep a deep, confident sleep.

Dawn had other plans.

I woke to a scene even worse than the day before. More blood. More bodies. More proof that whatever lived out there in the dark wasn't just hungry. It was relentless. Smarter than me. My defenses? My careful planning? My best efforts? Useless. The farm wasn't a fortress. It was a buffet.

Then came the trail-cam footage. Grainy, embarrassing proof of my incompetence. A mink, casually strutting around like a tiny criminal mastermind. A slick, slinking menace with dead eyes and endless audacity. It had tunneled under the coop, bypassing my traps like they were an insult. I watched it move with eerie precision, dismantling my efforts in real time. I had underestimated it. And now, my ducks had paid the price.

This wasn't over. Not by a long shot.

Further reinforcements to the duck house followed, driven by a newfound determination. The measures proved effective that spring. My remaining ducks were secure, and those that survived the ordeal gradually recovered. Many of them still roam the farm today, six years later. This pivotal experience taught me the harsh lesson that safeguarding my flock required more than hope and improvisation. It demanded a thoughtful, robust strategy to contend with the realities of rural life and the ever-present threat of predators.

The mink attacks weren't just brutal. They were a wake-up call. I had been lucky before. Too lucky. And now, luck had run out. My animals needed better. I needed to be better.

That's when I thought of my friend David. A fellow newbie farmer who raised sheep, David had recently brought on a pair of Maremma sheepdogs, massive white guardians bred for the sole purpose of keeping things like this from happening. I had seen them in action, calm, steady, unshakable, and I remembered the way they carried themselves, like nothing on earth could touch their flock. And nothing dared to try. That was the kind of presence my farm needed.

A lot of folks have never heard of the Maremma sheepdog, which is a shame, because they're not just dogs. They're legends. If you took a stroll through the rolling hills of Abruzzo or Tuscany a few centuries ago, you wouldn't just find them lounging in a sun-dappled piazza. No, they'd be out there, working, massive and unbothered, standing between a flock of sheep and whatever hungry beasts lurked in the shadows. These dogs weren't bred for companionship or blue ribbons. They were built by necessity, honed by the wolves they fought off, by the brutal terrain they patrolled, by the farmers who needed them more than they needed anything else.

And yet, their story didn't stop in Italy. Maremmas have since scattered across the world, proving that a good guardian is a good guardian, no matter the country, no matter the threat. These dogs aren't just farmhands. They've become frontline soldiers in nonlethal predator management, keeping balance in ecosystems where humans have done their best to screw things up. No fences, no traps, no bullets. Just instincts, centuries deep, doing what they were always meant to do.

Somewhere in my research, I stumbled onto a story that sealed the deal. In Australia, a team of Maremmas was deployed to protect a colony of endangered penguins from predatory foxes. And it worked. Not just for a week or a season, but for years. That story hit me hard. A dog, just a dog, could outdo all my obsessive planning and fencing. Maybe the lesson wasn't about having the perfect plan, but recognizing that often the best solution was simpler than I'd allowed myself to believe. It felt like a full circle moment. A reminder that sometimes, the oldest solutions are still the best.

By then, I was already half-convinced, but talking to David locked it in.

"You won't regret this," he told me over the phone one night. "These dogs are born for this. They'll watch over your ducks better than you could yourself."

And standing there, looking out at my farm, picturing those dogs moving like white ghosts through the fields, I knew he was right.

With renewed resolve, Allison and I agreed to adopt a Maremma puppy. We traveled to a farm in western Maine, a couple of hours from our own, where the potential new member of our farm family awaited.

The farm was quaint, reminiscent of our setup, but nestled amidst the rolling hills of Maine. Katie and Corey, the owners, greeted us warmly. They had spent years refining their skills on various farms across Italy and had recently settled back in the States. Their expertise with Maremmas was evident as they shared their experiences.

"We learned a lot about these dogs in Italy. They're not just protectors. They're part of the family," Katie explained as we walked through their farm, observing the sheep, ducks, and chickens coexisting peacefully.

We first met the puppies there amidst the gentle hum of farm life. One puppy, in particular, caught our eye. A quiet, somewhat reserved little fellow with an earnest gaze and incredibly soft fur, sitting quietly apart from his siblings. His eyes were brown pools of melancholy.

"He seems like he could be a good fit for you," Katie noted.

"What do you think?" I asked Allison, who had already fallen for the little one's melancholic expression.

"He's perfect," she replied, her voice filled with excitement and anticipation.

We decided then, watching the puppy interact gently with his siblings, that he would be ours. We named him Toby Dog. Or more specifically, we gave him the lofty name of Sir Bartleby de Mimsy-Porpington, Earl of Caledonia County. But we called him Toby Dog for short.

Livestock guardian dogs, LGDs if you want to sound like you're cool and you know what you're talking about, aren't *just* dogs. They aren't pets. They're workers, soldiers, legends, bred over thousands of years to do one thing: stand between their flock and whatever sharp-toothed menace thinks it's found an easy meal. Unlike herding dogs that corral livestock or hunting dogs that track or retrieve game, LGDs exist in a different category altogether. They are a blend of vigilance, brute strength, and an eerie, almost supernatural level of patience.

These dogs go way back to the first pastoral societies, when humans realized that protecting their animals meant protecting their own survival. Out of that necessity, entire breeds were forged: the Maremma in Italy, the Great Pyrenees in France and Spain, and the Anatolian Shepherd in Turkey, each one shaped by the landscape, the predators, and the farmers who needed them more than they needed sleep. These weren't dogs that chased threats away. They were dogs that stood their ground, dared the predator to make a move, and if it did—well, that was the predator's mistake.

To this day, that instinct remains. A good guardian dog knows the job before it's even asked. And that's not to say these dogs don't require training. It's just that they have instincts that make them predisposed to the job. They see the world in black and white: mine/not mine, safe/not safe. And they will defend their farms with their lives.

Livestock guardian dogs live with the flock, often from a very young age, which helps them bond with the animals they protect. This integration is crucial as it helps the dogs identify the livestock as part of their "pack."

Their protective behavior includes marking their territory by peeing on every tree and fence post they possibly can. This deters many predators from venturing close. They also use a variety of vocalizations to warn off intruders and to communicate with their charges and other guardians. Physical confrontation is a last resort. Their presence and warning barks are often enough to keep potential threats at bay. I call it marking and barking.

Daily, livestock guardian dogs patrol and mark the perimeter of the area where the livestock roam, constantly vigilant for signs of intrusion. They are particularly alert at night or when predator attack risk becomes likelier. The LGDs position themselves strategically, often on a rise or other vantage point, to watch over the animals. They rest intermittently but remain on high alert, ready to respond at the first sign of danger.

During an encounter with a predator, the guardian dog's first response is to alert the flock and the farmer by barking. If the predator does not retreat, the dog may engage in more aggressive displays, such as charging the intruder. Only as a last measure will they engage physically, ensuring to put themselves between the predator and the livestock.

The impact of having a livestock guardian dog on a farm is profound. Their presence can significantly reduce the loss of livestock to predators, which is a financial relief and decreases the emotional strain associated with such losses. Farmers can feel more secure, knowing their animals are being watched over day and night. Additionally, using LGDs is a humane and non-lethal method of predator management, aligning with modern wildlife conservation practices and ethical farming.

Bringing a livestock guardian dog onto the farm wasn't like adopting a lapdog from Petco. It meant accepting another living being who, frankly, would probably know more about farm security than I ever would. As they grow and bond with the livestock, their effectiveness as guardians only increases, making them invaluable farm family members. Their presence ensures not just the physical security of the farm animals but also contributes to the overall harmony and natural balance of the farm ecosystem.

Bringing a dog onto the farm made me think about privilege again. A farm dog, especially a specialized livestock guardian dog, isn't a cheap investment. It requires resources, planning, space, and security that not everyone has. A new puppy's impending arrival reminded me of the responsibility and good fortune embedded in my situation.

Toby's first home on our farm was the old duck house, the site of the mink's previous rampage. This choice was deliberate, meant to acclimate him to the area he needed to protect.

The training process was gradual, focusing on integration rather than obedience. I would sit with Toby in the duck house each morning, brushing his thick coat while we watched the sunrise.

"These are your birds now, buddy boy," I would whisper to him, hoping to instill a sense of guardianship in his young mind.

Our mornings continued with walks around the farm, Toby following at my side. "We need to check on everyone, make sure they're all okay," I explained to him, pointing out the ducks, the chickens, and the areas where predators had been spotted. During these walks, I observed Toby's instincts come to life. He was alert yet calm, his eyes scanning the surroundings with a protective vigilance.

A hawk circled overhead one morning while we were near the duck pond. Toby's reaction was immediate; his body tensed, and a low growl rumbled from his throat.

"Good boy, Toby," I praised. "That's your job, watch over them."

As weeks turned into months, Toby's presence became a constant reassurance. His integration with the ducks progressed smoothly, his natural instincts as a guardian overtaking the playful tendencies of his puppyhood. Watching him mature into his role, I felt a profound sense of relief and pride. I still look back on that time on the farm as one of my favorites. In fact, if you want to experience it for yourself, I ended up writing a *slightly* fictionalized account from Toby Dog's perspective in *Toby Dog of Gold Shaw Farm*, a novel for kids.

"Looks like he's taken to his duties quite naturally," I remarked to Allison one evening as we watched Toby patrol the perimeter of the duck house.

"Yes, he's exactly what we needed," she agreed, a smile spreading across her face as Toby settled down near the ducks, ever watchful, ever protective.

With Toby Dog's arrival, an essential piece to our farm's puzzle had been found. His diligent vigilance allowed our birds to roam freely without fear, and the incidence of predator attacks diminished significantly. Toby had become a guardian of our livestock and carved a special place in our hearts as a beloved family member. The situation with Toby Dog worked out so well that eventually, I decided to add a second dog. And in that choice, I'm pretty sure there's another lesson buried.

CHAPTER 7

BACKYARD DINOSAURS FOR FUN, PROFIT, AND BREAKFAST

I scooped up the mysterious white object lying in the grass. It was white, smooth, and roughly the size of a lemon. I held it gently like it might explode. Jemima Puddleduck watched me from a distance, her white feathers illuminated in the setting sun.

"Look at this thing!" I called to Allison, who was weeding the garden nearby.

"Wow, she really went for it, huh?" Allison laughed, wiping her hands on her jeans. "You gonna try it right now?"

"Absolutely," I said, practically bouncing inside with excitement. Minutes later, I was whisking the egg in a cast-iron skillet with an excessive amount of butter. The kitchen was filled with an aroma richer than any chicken egg had ever produced.

"Smells amazing," Allison said, peering over my shoulder skeptically. "You sure your stomach can handle that much butter?"

"It's not the butter that's gonna get me," I joked confidently. "It's duck or bust, baby!"

Later that night, as I crouched miserably on the cold tile floor of the bathroom, head spinning, stomach in revolt, Allison tapped softly on the door.

"You OK in there?" she asked gently.

"Define OK," I groaned between waves of nausea. "I had some gas station empanadas on the way home from work. That was a bad choice."

The plans for my first agricultural enterprise were always to raise ducks that would produce eggs. My brilliant business strategy envisioned happy ducks laying premium eggs I'd sell locally at upscale markets for a premium price. My farm would be so defined and distinguished by our egg-laying duck flock that we wouldn't need chickens. We would be known as The Farm Without Chickens. This was our brilliant farm plan. But brilliant farm plans usually end up blowing up spectacularly in your face.

Duck eggs are like the wagyu beef of the egg world. These little beauties pack a punch of rich, buttery flavor that puts chicken eggs to shame. They're also bigger, like softballs in the baseball game of chicken eggs. And for the world's bakers, duck eggs are the secret weapon in your pastry arsenal. Duck eggs are known for their high albumen content, making for a fluffier, more stable batter.

Bottom line: Duck eggs are the option for the discerning egg connoisseur. Trust me, once you try a duck egg, you'll never return to plain old chicken eggs again. Unless, of course, you're like me, a duck farmer allergic to duck eggs.

The second time I ate one of Jemima Puddleduck's eggs, the same thing happened again. And I started to realize that I had a serious problem on my hands. And since I was already making YouTube videos, I decided to document my experience and try it a third time. But this time, I was on a completely empty stomach, having nothing in my belly from the previous eighteen hours. I woke up first thing in the morning and scrambled a couple of duck eggs in some butter to see if that was what was doing it to me. A few short hours later, I had my answer. And I was one of the rare folks allergic to duck eggs.

Duck-egg allergies are not as common as chicken-egg allergies, but they can still occur. And to be clear, what I have is not technically an allergy, where you can have a range of symptoms, including hives, itching, swelling, difficulty breathing, and anaphylaxis. I have a duck egg intolerance, meaning I experience outstanding digestive symptoms such as diarrhea, vomiting, and stomach cramps after consuming duck eggs. It happens because my body can't digest a particular protein in the duck egg. The condition also extends to other waterfowl eggs, so goose eggs are off the table for me. Please note: *This is an extremely rare condition.*

Back in the day, when I dreamed of farm life back in DC, I always pictured myself surrounded by nature and raising my animals. The picture in my head included waking up every morning and eating eggs from my birds. But when I discovered I was allergic to duck eggs, my dream cracked open like an egg. I couldn't believe that something I had been looking forward to for so long was now out of reach.

I knew I had to find another source of eggs. And *that* is how I started raising chickens.

When it came to chickens, I was not exactly a fan. Their beaks are sharp, and their feet are pointy. Not nearly as pleasant as the roundness of a duck bill or the big flat surface of a goose foot. And chicken politics adhere strictly to Machiavellian principles. When I compare chickens to my ducks and geese, they have less personality, intelligence, and comedic relief. But with time and exposure, I have learned to appreciate them in my way.

Chickens are like the workhorses of the poultry world. They're reliable, they're efficient, and they're always ready to get the job done. They're the Ford F-150 of the bird world. They might not have the fancy features of a duck or goose, but they get the job done.

And let's not forget that chickens are like little dinosaurs. They are one of the closest living relatives to dinosaurs. I mean, look at them. They've got beady little eyes, sharp claws, and scaly feet. They're mini velociraptors, without the whole "trying to kill you" thing. They strut around the farm like they own the place, and honestly, they often do.

And let's not forget about their prehistoric vocalizations, the "bawk bawk" sound that makes you feel like you're on a safari in Jurassic Park.

It is a well-known fact that the origins of the chicken, the most respectable of feathered creatures, can be traced back to the wild jungles of the Far East. Indeed, it is said that the ancient ancestors of the chicken roamed the land in a state of feral savagery, a veritable force of nature, feared and revered in equal measure.

As time passed and the world changed, these wild birds slowly adapted to people's ways. They were tamed and domesticated, and their wild ferocity was replaced by domesticated ferocity.

Another remarkable aspect of chickens for your farm or homestead is that they are low maintenance. They're like the college roommates you never have to talk to. They're happy to do their own thing, and as long as you give them food and water and keep them safe from predators. That said, be ready for them to poop on the floor and take zero responsibility for cleaning it up. Also much like a roommate I once had in college.

"I can't believe I'm saying this," I muttered, poking through a box of fertilized chicken eggs from a friend, feeling distinctly unenthusiastic. "Chickens. How did it come to this?"

Allison leaned against the kitchen counter, sipping her tea. "Because duck eggs try to murder you?"

"Fair point," I conceded, carefully placing the eggs into the incubator one at a time, turning them gently as if they might detonate.

Days passed in a blur of anxiety until finally, soft peeps echoed from within the tiny shells. I crouched over the incubator, camera rolling.

"I can hear them!" I whispered, eyes wide with awe. "I'm about to be a chicken dad!"

Allison peeked inside the incubator and smirked.

"When do you think they'll come out?" I asked.

"I guess when they're ready," said Allison.

I couldn't wait to see them hatch. Tiny chicks emerging from their shells, all wet and fluffy. And as I waited for them to enter the world, I couldn't help but feel a sense of wonder and awe. Something so small and delicate can grow into something intense and resilient. The magic of hatching chicken eggs never gets old.

One by one, wet, fluffy chicks emerged into the world, tumbling clumsily in the warmth of the incubator. I couldn't help but grin like an idiot.

My first attempt at a chicken hatch was successful. Six of my eggs hatched, and those six chicks became the first chickens on our farm. Luckily for me, raising baby chicks is way easier than raising ducklings. They require the same resources, but a duckling makes infinitely more mess than a chick.

It was like having six tiny, fluffy babies, each with their personality. I spent hours sitting in the brooder, watching them as they grew. But as they grew bigger and stronger, I came to a bittersweet conclusion. I was excited to have my supply of eggs to eat, but four of the six hatched chicks were male. Because of this, I eventually had to dispatch three of them. Too many roosters in a flock can create violence and harm your hens. This is one of those inevitable truths of bird law. And bird law in this country is not governed by reason.

As the weather got colder, I had to decide which rooster would survive and which three would end up on the dinner table. Pablo Barn Cat supervised me on a chilly fall afternoon as I dispatched the three most aggressive cockerels. Each week for the subsequent three weeks, one of the cockerels would end up on our dinner plates.

Ultimately, we were left with Lucky the Rooster, Henrietta, and Marguerite, whom I affectionately nicknamed Margie. At first, things were OK, and the three made a nice little flock. Eventually, we noticed that Henrietta was getting beat up pretty badly. She had scratches and cuts all over her back, and many of her feathers were missing. Margie also had missing feathers but seemed in better shape than Henrietta. Mating between birds can often be vicious and violent, particularly with roosters and hens. You'll notice that the rooster's talons and spurs will wear off patches of feathers on the back of the head.

Lucky also developed a habit of attacking me each morning when I went to let the birds out. Like clockwork, I would be assailed from behind by a pile of angry feathers and razor-sharp spurs. I would have to grab him and hold him under my arm as I completed my chicken and goose chores each morning that winter. Between the overbreeding and the aggression, it was clear that Lucky had to go.

But even after I sent Lucky off to freezer camp, Henrietta the chicken was still not doing well. With Lucky gone, Margie was now at the top of the chicken pecking order, attacking Henrietta daily. The cuts on her back were getting worse, and one of them became infected.

"The poor baby! Why would Margie be so cruel to her?" asked Allison.

Allison has always had a soft spot for the weakest creatures on our farm. Anytime there's an animal injured, Allison's caretaker nature always kicks into high gear. And it was no different with Henrietta. Allison would clean and dress Henrietta's wounds, but Margie would attack again each day. I didn't want to separate the two chickens because I'd always heard that birds don't do well when alone, and they were my only two chickens on the farm. This went on for a week or two until I went out to check on the chickens one morning and found Henrietta dead in the coop.

I will never know whether Henrietta's death was the byproduct of infection, further aggression from Margie, or a little bit of both. But to this day, I know that her death was all my fault. My inexperience and poor preparedness led to the situation. It was an animal husbandry lesson I will carry with me until my dying day. That said, the experience also made me look at Margie differently.

Out of my original six chickens, I now had one lone survivor. Margie, who had effectively murdered her sister and outlasted her brothers, was a feathered Cersei Lannister. I started to think that she had engineered the death of all the other chickens so she could become the remaining lone chicken on the farm. Queen Margie the Wicked.

As I mentioned before, chickens are not meant to live alone. They are flock animals and need the company of their kind to feel secure. After the sad death of Henrietta, a few viewers of our YouTube channel generously offered us two Leghorn chickens, Blanche and Dottie.

We gladly welcomed them, and they quickly became a part of Margie's new flock. Later that spring, we introduced several more chickens to the farm.

Allison expressed her confusion. "I don't get it," she declared. "You always said you hated chickens. Now, you have almost as many chickens as ducks. Why?"

Her question stopped me. Hadn't I once sworn I'd never own chickens? Chickens were annoying. Chickens were messy. Chickens were, frankly, more trouble than they were worth. Or so I had thought. But somewhere along the way, my stance softened. Not all at once, not in some cinematic moment of poultry-induced enlightenment. Just little by little, until one day, I looked around and realized I wasn't just tolerating chickens. I was depending on them.

Because here's the thing: Chickens work. Sure, they lay eggs, but that's the least of it. Chickens do something ducks and geese can't. They dig. Their sharp little beaks, clawed feet, and relentless scratching at the earth like tiny archaeologists. They break up soil compaction, undoing the damage of heavier animals, aerating the land like it's their full-time job. And the insect control? Unreal. They hunt down flies and ticks with laser focus, making life infinitely better for our cattle. Every morning, I move the herd to fresh pasture, and two days later, the chickens follow, pecking through the cow pies, shredding manure, and devouring fly larvae before they even have a chance to become a problem. It's nature's cleanup crew in action. No chemicals, no machines, just the endless, hungry efficiency of chickens doing what chickens do.

And then, of course, there's the meat. Commercial chickens, especially the Cornish Cross variety, grow fast, too fast, raised in factory settings where they barely make it eight weeks before harvest. Our birds? Completely different story. We hatch our own flock's eggs. We keep the hens and butcher the roosters. It's slower, it's more intentional, and yeah, it makes for better, richer, more flavorful meat. There's something deeply satisfying about eating a bird that lived a real life, one spent scratching, pecking, being a chicken.

“Margie?” I called into the creeping darkness. Silence.

It was a warm August evening, and the usual bustle of the farm had quieted. I locked up the coop, counting birds as I went. Margie was missing. I flicked on my flashlight, sweeping it nervously across the tall grass. Toby Dog barked, alert and tense.

“Find her, Tobes,” I urged quietly. He darted forward, sniffing intensely. Minutes passed as we combed through the orchard.

Finally, Toby stopped abruptly by a swale, whining softly. I approached cautiously and saw white feathers scattered among the grass, illuminated ominously by my flashlight beam.

“Oh no,” I whispered. Margie’s body lay silent and still, her queenly reign violently ended.

All of the physical evidence suggested she was likely taken by a hawk or owl. This incident had marked the only predator loss since Toby had matured into his role as a livestock guardian dog. Ultimately, Margie’s reign ended just as dramatically as she lived, a stark reminder of farm life’s cyclic and often harsh realities.

Later, sitting at the kitchen table with my head in my hands, Allison rubbed my back soothingly. “You did everything you could.”

“Maybe,” I sighed. “But part of me wonders if Margie’s ghost will come back to haunt me.”

She squeezed my shoulder, smiling faintly. “If any chicken could pull off a haunting, it’d be her.”

CHAPTER 8
NIGHTMARE DEMONS OR THE MOST SUSTAINABLE FORM OF POULTRY?

The sun dipped below the hills as we left Salamanca, casting long streaks of gold across the parched fields of Spain's Extremadura region. Months of dead-end leads, ambiguous email messages, and one transatlantic flight all pointed to this moment: finding Eduardo Sousa, the legendary goose farmer. And now, as I pulled up the driveway, doubt crept in. What if it was all for nothing?

For me, geese had always been an enigma. Birds with a reputation as temperamental yard dragons but capable of profound environmental harmony. They seemed to straddle two worlds: the chaotic hiss and honk of barnyard chaos and the quiet, methodical work of trimming pastures with all the calm of a Zen gardener. Allison, who sat in the passenger seat, thumbing through a Lonely Planet guidebook, says they remind her of robot dinosaurs.

What I'd read about Eduardo Sousa suggested something even more profound: Geese could challenge the very foundations of modern farming, offering a sustainable, ethical alternative to some of agriculture's most controversial practices. If his claims were true, this wasn't just a story about foie gras. It was a story about possibility. About how one farmer and a flock of birds could reimagine what it means to live in balance with the land.

"This is it," I said, mostly to myself. "This is where Eduardo Sousa lives." The name felt heavy on my tongue, like speaking it aloud might summon him. Months of research, endless emails that had gone unanswered, and now a transatlantic flight had brought me here, chasing a story that felt too good to be true.

Ethical foie gras.

Foie gras. Just saying the words feels like you're summoning a tuxedoed waiter to your side, napkin over arm, platter in hand. It's the kind of dish that whispers decadence in one breath and shouts outrage in the next. A silky, velvet-rich indulgence born from the liver of ducks or geese, coaxed into unnatural largesse by gavage, a practice as barbaric as it sounds.

Picture it: a metallic tube, jamming that tube down the throat of the bird, which is then force-fed until it is left waddling under its own surreal proportions. For weeks, it eats. And waits. And endures. Animal-rights advocates rightly decry gavage as cruel and unnatural, pointing to images of confined birds, their movements restricted, their lives reduced to a narrow cycle of eating and waiting. Governments in some countries have banned the procedure outright, while others continue to defend the practice as culinary tradition.

Yet amidst the din of foie gras–related protests and Michelin-starred rationalizations, enter Eduardo Sousa, a man as unlikely as his claim: foie gras without force-feeding. No tubes, no cages, just geese wandering the open land like feathery epicureans, grazing their livers swollen. It sounded too good to be true, the kind of story Hallmark might balk at for being a bit much. But there he stood, a doughy Don Quixote of liver pâté, quietly insisting that freedom, and maybe a sprinkle of joy, was the secret ingredient. It was a claim that demanded a closer look, and I wasn't about to let it slip away.

I'd first heard of Eduardo Sousa when I read Dan Barber's *The Third Plate* during my earliest days of going down the regenerative agriculture research rabbit hole. The book paints Sousa as more than a farmer, a kind of gastronomic poet who rewrote the rules of foie gras with nothing but an open field and a deep respect for the geese under his care. Barber called him a visionary, a pioneer who dared to imagine a world where nature and indulgence weren't at odds. But it wasn't just Barber's words that hooked me on the legend of Eduardo Sousa. It was the video and photographs I would soon find on the internet. Geese wandering freely under the sprawling shade of ancient oak trees, acorns scattered like nature's buffet at their feet. Olive branches swollen with green orbs just waiting to be shredded by serrated goose bills. Articles described Eduardo as a revolutionary, a man who had ruffled the feathers of industrial producers simply by showing that their methods weren't the only way. In 2006, his farm, La Patería de Sousa, was awarded the prestigious Coup de Coeur Award at the Salon International de l'Alimentation in Paris. This accolade is particularly noteworthy as it is traditionally dominated by French producers who utilize conventional methods involving gavage.

For someone like me, trying to reimagine what farming could look like on a patch of Vermont soil, this was goose-feathered catnip. I wanted to expand our operations, to root them in the wisdom of European traditions without losing sight of ethics or sustainability. Eduardo's story was proof that it could be done. Physical evidence that luxury didn't have to come at the cost of cruelty. So I spent hours poring over every article, every sliver of information, trying to understand how he had managed to turn an industry built on force into one built on trust.

At first, it was just curiosity, an intellectual fascination with a farming technique I'd never heard of. But over time, it turned into something else. A question had taken root in my mind, growing into a relentless itch: Could we do this back home? Could we make foie gras without compromising our values?

Allison had been skeptical at first. "We're really going to fly all the way to Spain for this?" she'd asked over breakfast one morning, her eyebrows once again raised as she buttered her gluten-free toast.

"Yes," I'd replied without hesitation. "If this works, it could change everything."

And so here we were, navigating the winding roads of rural Spain, a place that felt as timeless as the story I was chasing. But finding Eduardo wasn't going to be easy. He was notoriously elusive. Despite his fame in farming circles, Eduardo didn't advertise. There were no flashy social media posts, no marketing campaigns. Just a small shop in a tiny town where he supposedly sold his pâtés and charcuterie.

We'd landed in Madrid two days ago, and so far, every lead had ended in frustration. Emails had gone unanswered. Phone numbers were duds. And now, as we approached the shop where I'd hoped to finally catch a glimpse of the man himself, my heart sank. The shutters were closed, the windows dark.

I parked the overly small European rental car and stepped out, the heat of the day still radiating off the cobblestones. "It's closed," I said, my voice tight with disappointment.

Allison climbed out beside me, squinting at the shop. "Closed? But it's not even siesta time. Are you sure this is the right place?"

"Positive," I said, pointing to the faded wooden sign above the door. La Patería de Sousa.

I glanced at my watch. It wasn't even five yet. According to Google, the shop was supposed to be open until eight. I knocked on the door, hoping against hope that someone might answer. Nothing. Just the faint rustle of leaves in the breeze.

"What now?" Allison asked, her tone edging into irritation.

I sighed and leaned against the car. The weight of the journey, the anticipation, and now the uncertainty felt like too much all at once. "I don't know," I admitted. "I thought this would be the easy part."

Allison crossed her arms, her expression softening. "Hey, it's not over yet. We'll figure it out."

I wanted to believe her. But as I stood there, staring at the closed shop, the dream of meeting Eduardo Sousa, of learning his secrets and bringing them back to Vermont, felt like it was slipping through my fingers.

Still, I wasn't ready to give up. Not yet.

As my search for Eduardo continued, I found myself thinking more broadly about geese, not just as animals or farmhands, but as participants in the history of agriculture. What made geese unique? Why had they been cast aside in favor of faster-growing, grain-dependent birds like chickens? If Eduardo's geese represented a new path forward, perhaps understanding their past could shed light on what we'd lost. And what we stood to regain.

Goose farming is one of the oldest forms of domesticated poultry management, tracing back thousands of years to ancient Egypt. Picture the banks of the Nile thousands of years ago, where geese waddled under the watchful eyes of Egyptian farmers. Back then, these birds were more than just a source of meat, eggs, and feathers. They were prosperity incarnate, symbols of abundance in a world that often teetered on scarcity.

The Romans took it a step further, practically deifying their geese. Legend has it that a gaggle saved the Eternal City from a Gallic invasion, honking their alarm in the dead of night as enemy troops crept toward the Capitoline Hill. Across medieval Europe, geese became the backbone of countless farms, thriving in places where harsher climates and barren lands left other livestock wanting. By the eighteenth century, they were practically celebrities in agricultural communities, prized for their golden fat, sumptuous meat, and feathers soft enough to cradle dreams. Holiday feasts weren't complete without a goose at the table, its crisped skin glistening like a crown jewel.

But history, as always, marched forward. The industrial age brought relentless focus on efficiency, favoring fast-growing, grain-fed poultry like chickens and turkeys. Geese, with their slower growth and sprawling size, were deemed impractical for the tightly packed cages and assembly lines of modern farming. In America, the shift was even more stark. Though early settlers brought geese with them, the abundance of cheap corn and a cultural preference for convenience quickly elevated the chicken to national icon status, leaving geese to languish as an afterthought, their once-proud place in the barnyard reduced to the odd Christmas roast.

And yet, geese remain an unsung hero of sustainable farming. Unlike their grain-fed cousins, they thrive on grass and weeds, grazing like miniature cows in feathered suits. Their role as nature's lawnmowers reduces the need for heavy machinery, while their manure enriches the soil, weaving them into the fabric of regenerative agriculture. Their diet, a mix of grass, weeds, and other vegetation, avoids the ecological pitfalls of growing feed crops like corn, which guzzle water, drain nutrients from the soil, and contribute to greenhouse-gas emissions through fertilizers and transport.

In a world grappling with climate change and unsustainable farming practices, geese present a radical simplicity. They require little, give much, and tread lightly. Every honk, every nibble at a weed, every bit of fertilizer they leave behind tells a story of balance. A story we need more than ever.

On our Vermont farm, I see it every day. There's a patch of land where the geese graze, their beaks busy among the grasses and wildflowers. A lone dandelion stands no chance, nor does the invasive ragweed encroaching on the pasture's edge. In their wake, the field grows greener, richer, and teeming with life. It's as if they know they're part of something larger, an ecosystem humming with quiet symphonies. And at dusk, as the sun sinks low and the geese gather in soft murmurs, I'm reminded why I'll always say it: Geese aren't just good for the planet. They're good for the soul.

The Airbnb we'd rented in Extremadura was a rustic little apartment in the center of a village that felt like it hadn't changed in centuries. The walls were whitewashed, the tiles on the roof were cracked with age, and the windows looked out over a landscape that seemed to stretch forever. Rolling fields dotted with ancient oaks, their branches heavy with acorns. It was beautiful, but as I paced the kitchen that evening, the scenery felt more like a mocking backdrop to my frustration.

"Maybe he doesn't need the shop," said Allison. "If his foie gras is as famous as everyone says, he's probably got a waiting list a mile long. He doesn't have to sit around for walk-ins."

That made sense, but it didn't help. "It just feels like I'm missing something," I muttered, slumping into a chair across from her. I pulled out my phone and stared at the email thread where I'd tried (and failed) to connect with Eduardo over the past few months. Nothing but my own words stared back at me.

Allison took a sip of her tea and studied me. "What's the plan?"

I rubbed my temples, the weight of the trip pressing down on me. "I'll go back to the shop first thing tomorrow. If it's still closed…" I trailed off, not wanting to say the words out loud. If the shop didn't pan out, I didn't have a Plan B. But I've spent most of my life winging it, so why stop now?

Allison studied me quietly over the rim of her tea mug. "And if you still can't find him?"

I sighed, letting the question hang for a moment. "Honestly? No clue. Maybe I'll just wander the streets of Extremadura yelling his name."

She cracked a smile. "Well, at least you'd make the local news."

I laughed, feeling some of the tension ease. "Thanks for sticking with me through all this."

She shrugged lightly, hiding her own smile. "Someone has to keep you out of a foreign prison."

The next morning, I woke with renewed determination. The air was crisp and cool as I stepped outside, the kind of morning that makes you feel like anything is possible. Antonio, our Airbnb host, was tending to the small garden in the backyard. He waved me over with a knowing look.

"You find Eduardo?" he asked, his Spanish thick and melodic.

"No," I stuttered in my nearly incomprehensible Spanish. His shop had been closed. *"Su tienda, uh, estaba cerrada."*

Antonio frowned. He gestured for me to follow him, leading me to a small patio shaded by grapevines. He pulled out a piece of paper and sketched a crude map of the area. His finger landed on a spot near the edge of town. *"Finca,"* he said.

A farm? My heart jumped. Did he know where it was? *"¿Sabes dónde está?"*

"Sí," he said with a shrug, as if it were the most obvious thing in the world.

I stared at the map, barely daring to believe it. *"Gracias, Antonio. Muchas gracias."*

He smiled and waved a dismissive hand. *"De nada. Pero cuidado,"* he warned, pointing at me with a serious expression. *"Eduardo no siempre recibe visitas."*

I nodded knowingly, even though I couldn't make out his final words. My Spanish just wasn't there.

The drive to the farm felt surreal. The roads became narrower and more rugged, winding through groves of oak trees that seemed to whisper secrets in the breeze. Ennio Morricone's "Ecstasy of Gold" played on loop on the rental's Bluetooth. The GPS chimed occasionally, but it felt like I was being guided by something older than technology. Purpose was pulling me forward.

Then, as I crested a hill, I saw it: a vast expanse of land, dotted oak trees. A pig farmer wielding a crook herded his black Iberians through the meadows, preventing them from staying too long in one spot and over-grazing. I parked at the edge of the property and stepped out. For a moment, I just stood there, taking it all in. The grasses, the trees…the sheer simplicity of the place. It felt like I was adventuring.

But now came the hardest part: finding Eduardo and convincing him to let me see his operation. The pig-herding farmer had told me to follow the *camino* down to the farmhouse. Walking down the dirt path, I felt excitement and nerves bubbling in my chest. The journey was far from over, and I knew the real challenge was just beginning. The sharp caw of a distant crow, the rhythmic clatter of leaves stirred by the breeze, and the occasional low honk from a goose waddling nearby. This was no ordinary farm.

Ahead, a large metal gate blocked entry to the farmhouse yard, its iron bars weathered and slightly overgrown with ivy. Beyond it, the whitewashed house stood stoically, its simplicity almost daring me to approach. I couldn't see a soul. No sign of Eduardo, no workers, no vehicles. Just the geese and the land, stretching endlessly into the horizon. It was as if the farm existed in its own world, untouched by time or the noise of the outside.

I lingered at the gate, shifting my weight from one foot to the other, unsure what to do. There was no doorbell, no sign of how to announce myself, only the stillness of the place. I checked my watch, waited a few more minutes, then reluctantly turned to leave. I couldn't shake the growing sense that I was intruding on something sacred.

Just as I reached the edge of the lane, the clanging of metal drew my attention. I turned to see an older man with a wide-brimmed hat and a deliberate gait walking up to the gate. He unlatched it and opened it wide, his movements unhurried but purposeful.

Seeing him, I froze for a moment before quickly running back. "Eduardo?" I called out, half-expecting him to shoo me away.

He looked up, his face deeply lined and tanned from years under the sun, and regarded me with a mixture of curiosity and caution.

"Eduardo Sousa?" I ventured, stepping toward him.

He didn't answer right away. Instead, he tilted his head slightly, studying me like one of his geese might inspect a stranger in their flock. Finally, he spoke, his Spanish slow and deliberate. *"¿Quién eres?"*

"My name is Morgan," I began, fumbling to piece together the rest of my explanation in Spanish. *"Yo, uh, vino de Estados Unidos, para aprender sobre tu granja. Yo también soy criador de gansos."*

For a moment, he simply stared at me, not saying a word.

I had spent weeks imagining what it would be like to meet Eduardo Sousa. I pictured us walking through his fields, him pointing out the subtle differences in oak trees that bore the best acorns, and me scribbling furiously in my notebook. I thought we'd share stories about our farms, and he'd laugh when I told him about the chaos of raising geese in Vermont winters. But now, those images felt like they were dissolving into the night.

Looking back, I've come to realize just how intrusive it is to show up unannounced at a farmer's home. At the time, I thought of myself simply as a curious visitor, seeking inspiration. But now, after years of running my own farm, I understand how deeply disruptive those unplanned visits can feel. As Gold Shaw Farm got internet-famous, we began seeing up to twenty unannounced visitors a week. Countless strangers pulling up the driveway to snap pictures, see the animals, or shake my hand. Most of these folks are exceptionally nice and well-meaning, but the visits drive me insane. Farming isn't just work. It's a life of rhythms and routines, and those rhythms are easily unsettled by unexpected guests. More importantly, this is our home, and having people enter without notice feels like a violation of something personal and sacred.

It's with that perspective that I now reflect on my visit to Eduardo Sousa's farm. I was a guest in his country, a stranger at his gate, and my determination to meet him overshadowed the respect he deserved. I saw my enthusiasm as harmless, but in hindsight, it was an uninvited intrusion into his world. Writing about this episode fills me with regret and embarrassment. It's a hard lesson, but one I've learned firsthand. To Eduardo and to any farmer whose space I disrupted, I hope this serves as the apology I never gave in person. And if you're reading this with thoughts of emulating my behavior: *Don't do what Donnie Don't does.*

Eduardo led me around the side of the house and onto a dirt path that wound through the heart of the farm. It was clear he didn't speak much English, but he didn't seem to mind my broken Spanish either. Words felt almost unnecessary here.

We walked through the pastures where dozens of geese roamed and ripped grass from the ground. Eduardo leaned on a tree, his eyes scanning the birds. He began to speak, his voice low and steady, and I scrambled to catch his words.

"Las ocas saben qué hacer," he said, gesturing to the geese. *"No necesitan forzar nada."*

The geese know what to do. They don't need to be forced.

I watched as one particularly large goose plucked an olive from the ground. Others joined in, their movements unhurried, almost meditative, as if savoring the season's bounty. Eduardo pointed to the olive trees, their branches heavy with fruit.

"Aceitunas," he said, his voice calm but reverent. *"Dan la comida. Lo natural."*

The olives. They give the food. It's natural.

He knelt to pick up a freshly fallen olive, its surface smooth and rich with oil, and held it out to me. I turned it over in my fingers, its faint aroma hinting at the flavors it would bring to the geese.

"Las ocas engordan solas," he explained, *"porque están felices."*

The geese fatten themselves because they're happy.

It was such a simple statement, yet it felt profound. This wasn't just farming. It was philosophy.

As we continued walking, Eduardo shared more about his methods. The land was meticulously maintained, not in the clinical, industrial sense, but as a living, breathing ecosystem. Wildflowers grew between the trees, attracting pollinators. The geese roamed freely, their movements shaping the landscape as much as the landscape shaped them.

Occasionally, Eduardo would stop to show me something: a cluster of herbs growing at the base of an oak, a patch of soil where the acorns were most abundant, the skeletal remains of an old cart that seemed to have been there for decades. Each time, he spoke with reverence, as if every detail was essential to the farm's balance.

By the time we circled back to the house, the sun was high in the sky, and I felt as though I'd glimpsed a way of farming that was almost otherworldly. Eduardo paused at the door, his hand resting on the frame, and looked at me.

"Es un trabajo de generaciones," he said. *"No mío solamente."*

It's the work of generations, not just mine.

I nodded, pretending to understand what he meant.

When he handed me his contact information and motioned toward the path, I knew our time together was ending. But as I made my way back to the car, the beauty of his farm stayed, and the weight of his words stayed with me. Or at least they stayed with me after I got home and fed the conversation into Google Translate.

Producing ethical foie gras is my agricultural white whale. For six years now, every goose-butchering week, I've approached the task with cautious optimism, hoping to discover livers so plump and golden they'd make a Michelin-starred chef weep. And yet, year after year, my geese deliver perfectly respectable, utterly ordinary livers. No yellow. No buttery richness. Just regular old goose livers, as if they didn't understand the assignment.

It'd be easy to call my whole ethical foie gras endeavor a bust. To look at every normal goose liver and feel like the mark had been missed. But farming doesn't hand out neat little endings tied up with bows. It's messier than that. Every season feels like part of a never-ending back-and-forth. A conversation between you and the animals, the land, your big plans, and the reality that maybe it rained too much, or not enough.

"A conversation with the land?" you may ask. "What's next, a debate with the barn? Negotiations with the scarecrow?"

And you're right. This book isn't supposed be *Zen and the Art of Goose Liver Maintenance*. You're probably wondering: What's the actual reason that I haven't been able to reproduce Eduardo Sousa's results? Well, I do have a hypothesis. The issue, I suspect, lies in my geography and climate. Eduardo is farming in a far warmer climate. His geese feast on a much longer growing season that can produce calorie bombs like acorns, olives, and high-octane lupin plants that simply don't thrive in my colder Vermont soil. Meanwhile, my orchard of apple, mulberry, and chestnut trees are still finding their footing, offering my geese a steady but modest diet. But it's like trying to bulk up on kale instead of pizza. It just doesn't hit the same.

Still, there's no bitterness here, only acceptance. Raising geese, even without foie gras in the picture, has become one of the most fulfilling parts of my farming life. I've come to believe that the world would be a far better place if more people were raising geese and serving goose meat at their tables.

There's that old saying: Aim for the stars, and you'll land on the moon. Well, in my world, it's more like: Aim for foie gras, and you'll land on a deliciously sustainable Christmas dinner.

ALLISON AND ME IN FRONT OF OUR ROW HOUSE IN CAPITOL HILL.

MY FIRST GARDEN BEDS IN DC. ALSO PICTURED, MY FIRST GOAT.

AN AERIAL PHOTO OF THE PERMACULTURE ORCHARD ROUGHLY NINE MONTHS AFTER PLANTING.

I TOOK THIS PICTURE THE DAY WE CLOSED ON THE FARM BACK IN JULY 2016.

SWALES AND BERMS WITH BABY TREES IN THIS PHOTO OF THE ORCHARD BACK IN 2018.

ALLISON AND PABLO BARN CAT, THE FIRST TWO PERMANENT RESIDENTS ON GOLD SHAW FARM.

I SPROUT ROUGHLY 500 CHESTNUT TREES A YEAR. HERE'S A FRESH BATCH READY FOR THEIR YEAR ONE PLANTING.

MARGIE THE MURDER CHICKEN AND HER SIBLINGS SHORTLY AFTER HATCHING.

BARN AT SUNRISE. THE KIND OF GLOW YOU ONLY GET FROM UNPAID REPAIRS.

PABLO BARN CAT AND LIL BARN CAT DURING HER "OUT-DOOR CAT" ERA. THEY WERE THE BEST OF FRIENDS.

ALLISON AND GRETA THE DUCK SHORTLY AFTER THE HORRIFIC MINK ATTACK OF 2019.

BABY TOBY DOG, SHOWING UP FOR HIS FIRST DAY OF WORK ON THE FARM BACK IN NOVEMBER 2019.

HE'S NO LONGER A BABY, BUT TOBY DOG IS STILL A BIG BABY. HE HATES LEAVING THE FARM AND I HAVE TO PICK HIM UP TO PUT HIM IN THE TRUCK WHEN I NEED TO TAKE HIM TO THE VET OR THE GROOMER.

BUT HE REMAINS MY BEST FRIEND.

ALLISON LOVES TENDING TO THE BABY BIRDS. SHE'S HOLDING RON SWANSON, THE DUCK WHO THINKS SHE IS A GOOSE.

TOBY DOG IS THE OFFICIAL GOSLING INSPECTOR OF THE FARM.

THIS ONE TIME, THE GOSLINGS ESCAPED AND TRIED TO MAKE A RUN NORTH FOR THE BORDER.

A MAJESTIC FLOCK OF GEESE WORKING THE PASTURE.

PABLO BARN CAT HAS A FEROCIOUS YAWN.

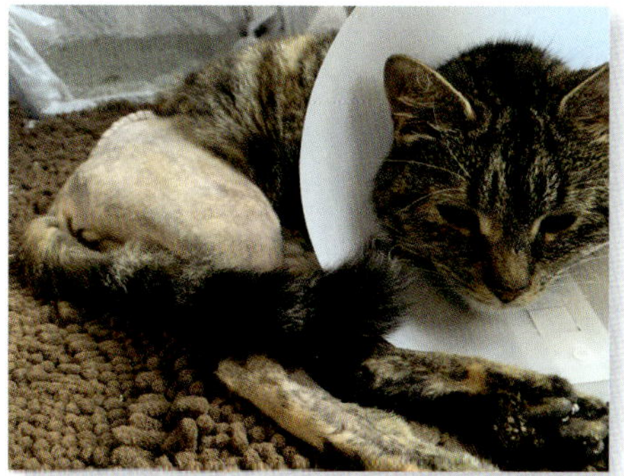

LIL BARN CAT SHORTLY AFTER HER ACCIDENT. THE VET HAD TO REMOVE HER TROUSERS.

THESE DAYS, LIL BARN CAT IS A FULL-TIME HOUSE CAT AND IS MY CONSTANT OFFICE COMPANION.

TOBY DOG REPRIMANDS ABBY DOG FOR HER POOR BEHAVIOR.

ABBY DOG EYES A WHITE CHICKEN.

INSPECTING EGGS WITH ABBY DOG. IN THE FOREGROUND, YOU CAN SEE A GOOSE EGG, A DUCK EGG, A REGULAR CHICKEN EGG, AND A SILKIE CHICKEN EGG.

GINNY BARN CAT AND HER MOTHER, MOLLY BARN CAT, SHORTLY AFTER ARRIVING AT THE FARM IN 2021.

MY BELOVED MOLLY MURDER MITTENS MAY BE GONE, BUT SHE'LL NEVER BE FORGOTTEN.

GINNY BARN CAT LOVES STALKING FROM THE FENCE POSTS.

WHENEVER I HUG ONE OF THE DOGS, LIFE FEELS SLIGHTLY LESS BROKEN.

ANNE OF GREEN GABLES ON THE DAY SHE GAVE BIRTH TO OUR FARM'S FIRST CALF, BELINDA CARLISLE. THIS PHOTO WAS TAKEN IN MARCH 2022.

BELINDA CARLISLE WAS A SASSY HEIFER.

BELINDA CARLISLE HAS BECOME A VERY ATTENTIVE MOTHER. HERE SHE IS WITH HER FIRST CALF, CYNDI LAUPER, IN SPRING 2025.

HERE I AM BRUSHING MACHO MAN RANDY SAVAGE, OUR FARM'S GENTLE BULL.

THERE ARE FEW THINGS MORE ADORABLE THAN SCOTTISH HIGHLAND CALVES MUNCHING ON FRESH GRASS.

TOBY DOG SITS ON DUTY DURING AN EPIC SUNSET.

IF YOU TOLD ME TEN YEARS AGO THAT THIS WOULD BE A TOTALLY NORMAL MORNING FOR ME, I WOULD HAVE SAID YOU WERE INSANE.

BARNACUS, TELEMACHUS, AND LARN, OUR FARM'S GOATS. (SHOOT, I NEVER EVEN TOLD YOU GUYS ABOUT MY CRAZY GOAT STORY!)

CHAPTER 9

I AM NOT A CAT PERSON

I am not a cat person.

People don't believe me when I say that. They hear that I've owned as many as four cats at one time and laugh. *You must be joking?* I'm not.

"Yes, yes," people say, rolling their eyes. "You're totally not a cat person. Weirdo."

I get it. On paper, it doesn't add up. But hear me out: Cats and I have always had an understanding. They're independent. I respect that. They don't need constant validation, and neither do I. They go their way. I go mine. It's a professional arrangement, not a love affair. Also, I'm wildly allergic. Cats make me sneeze like I'm snorting an entire pepper shaker. My eyes swell up. My throat itches. It's a whole thing.

Dogs, on the other hand? Dogs are simple. They're loyal. They love you with an unwavering devotion that fills every room they're in. Dogs are my people.

But look, this isn't some Disney origin story where I end up adopting a tuxedo kitten and learning the true meaning of love. No, this is logistics. Cats and I have a professional relationship. And that's it. When Allison and I started a poultry farm, I knew cats must be part of the deal. Chickens and ducks, and more specifically, chicken and duck feed, attract rodents. Rodents bring disease, destruction, and chaos. Barn cats are the farm's natural pest control. It's a business arrangement, plain and simple. I wasn't adopting pets. I was hiring employees.

It started on a crisp May evening, where the Vermont air feels alive with promise, but the ground still holds onto the disgusting chill of mud season. It had been two weeks since I finally moved to the farm full-time and started my new job, freshly escaping from the buttoned-up grind of DC.

I was scrubbing the day's filth off my hands with a vigor that would make Howard Hughes proud, when Allison lobbed her grenade of a question that would upend my carefully calibrated system of barn cat management.

"What do you think about getting another cat?"

I paused mid-scrub and turned to look at her. Pablo Barn Cat had already become her constant shadow. She loved that cat more than she loved me. And now she looked like she had planned this ambush.

"Another cat?" I repeated.

"Nick and Zia have a litter of kittens," she said, setting her mug down. "They're trying to find homes for them."

Our friends, Nick and Zia had moved to Vermont around the same time as us, chasing a similar city-life-to-rural-life dream. But the vacant house next door to their rental came with an unexpected bonus: feral cats. Lots of them. The most infamous was a flabby tabby they called Big Edie. She wasn't their cat. She belonged to no one. But she'd taken up residence in the abandoned house next door. She and a second cat (Little Big Edie) got trapped inside the house. When Nick and Zia liberated the cats from the abandoned house, they realized she was pregnant. Big Edie gave birth to a litter of five tiny furballs. Nick and Zia did their best to care for them. (Side note: for the love of Bob Barker, spay and neuter your pets, people!)

"And why, exactly, do we need another cat?" I asked, drying my hands on a dish towel.

"Pablo's great," she began, her voice slipping into that patient tone she used when trying to talk me into something. "But I think it would be really good for him to have a friend."

"Pablo doesn't need a buddy," I said, shaking my head. "He barely tolerates me, let alone another cat."

"It's not just about Pablo," she said, her expression softening in a way that made me deeply suspicious. "Another cat would be good for the farm. More mice taken care of, fewer rodents in the feed. It's practical."

Practical. Right. Like installing a disco ball in the barn.

I leaned against the counter, arms crossed. "Allison, we're about to get ducks. Forty ducks. That's forty loud, messy reasons to question every life decision I've ever made. Do we really need to throw a kitten into the mix?"

"It wouldn't be that much work," she said, all innocence. "I'd handle everything. You wouldn't even have to lift a finger."

I gave her a look.

She smiled, the kind of smile that crinkled her eyes just enough to make me question whether I was the villain here. "You're just mad because Pablo chose me over you."

"It's not a competition," I muttered, though we both knew it absolutely was.

Her smile faltered, and she looked down at her tea. "You know how lonely I was this past winter," she said softly. "How much Pablo helped."

The air between us shifted, her words cutting through my resistance like the first real thaw after a brutal freeze. I hated thinking about that winter. Her voice on the phone so small, the farmhouse silence pressing down on her like a weight.

"This isn't just about rodent control," she said. "Another cat would make the farm feel…more alive."

I studied her face, lit softly by the kitchen light. She wasn't just asking for a kitten. She was asking for connection. For life. For something to keep the silence at bay.

"Fine," I said, dragging a hand through my hair. "But it stays outside. Like Pablo."

Her face lit up, and she grabbed her phone, already texting Zia. "You're going to love this kitten," she said, her excitement spilling over.

"Don't push it," I muttered, yanking on my boots to head outside.

And there'd be a kitten.

"But this doesn't make me a cat person."

It was barely light out when I heard the crunch of tires on the gravel driveway. I glanced at the clock: 5:07 a.m. The coffee in my mug was lukewarm, the farmhouse silent except for the faint creak of the floorboards beneath my feet. Allison was still asleep upstairs, the curtains drawn against the pale gray dawn.

I stepped outside, pulling my jacket tighter against the early morning chill. Nick's car idled in the driveway, its headlights cutting through the dawn mist.

Nick hopped out, dressed in his breakfast-chef whites, looking far too chipper for an hour that still felt morally wrong to be conscious. He was holding a small plastic crate under one arm. "Morning," he said cheerfully.

"Morning," I replied, my voice still rough from sleep.

"Here she is," he said, holding the crate out to me. Inside, a tiny tabby kitten huddled in the corner, her body pressed tight against the plastic as if trying to disappear into it. Her fur was a patchwork of gray and brown, with faint orange streaks that caught the light when she shifted.

"She's all yours now," Nick said, giving me a grin before heading back to his vehicle. "Good luck."

And just like that, he was gone, leaving me standing in the driveway, holding this trembling little creature who had no idea her world had just been upended.

I carried the kitten into the mudroom, setting her down in the corner we'd prepared the night before. There was a small bed made from an old blanket, a bowl of water, and a kitten food dish. Everything a kitten could need. At least, that's what I told myself.

"OK," I said softly, crouching down to peer inside. "This is your new place, kid."

She stayed put, her wide eyes fixed on me with a mix of fear and uncertainty. Her tiny body trembled, and for a moment, I thought she might never move again. She emitted a soft mewl of fear.

I sighed, sitting cross-legged on the cold mudroom floor. "Look, I get it," I said, more to myself than to her. "This isn't exactly what I wanted either."

For a while, we just sat there, the silence broken only by the faint hum of the mudroom freezer and the occasional creak of the house settling. I reached out slowly, picking up a piece of straw from the floor. I held it out, letting it dangle in front of her crate.

She didn't move at first, her eyes darting between me and the straw. But then, with a hesitant twitch of her paw, she swiped at it.

"There you go," I murmured, moving the straw gently back and forth.

Her swipes grew quicker, more confident, her tiny claws catching the straw and pulling it toward her. She batted it back to me, her movements awkward but deliberate, as if she understood this was a game.

I chuckled softly and flicked the straw across the floor. She pounced on it with surprising speed, her tail flicking in tiny, triumphant jerks. For a moment, I forgot about everything else—the early hour, the cold floor, the fact that I hadn't asked for another cat.

Before she had arrived, Allison and I had agreed to name her "Lil Barn Cat" because she was a barn cat who would be smaller than Pablo Barn Cat. To this day, on our farm, all cats carry the surname of "Barn Cat," just like the dogs carry the surname "Dog."

We played like that for a while, the straw darting and Lil chasing, her tiny frame a blur of movement. She tumbled and leapt, her cautious demeanor giving way to pure, unfiltered playfulness.

Eventually, she flopped onto her side, the straw pinned beneath her paw, her chest heaving with effort. I leaned back against the wall, watching her catch her breath.

"Relax, champ. The barn-league scouts don't get here until next month." I quipped.

Her eyes met mine, her gaze softer now, less wary. She batted the straw one last time in my direction before curling up in the bed we'd made for her. Her small body rose and fell with the steady rhythm of sleep, her earlier trembling replaced by a quiet calm.

But I was not a cat person.

Training a barn cat might seem daunting to those unfamiliar with their role, but the process is relatively simple and rewarding when done correctly. You'll need a large crate, similar to those used for dog training. The cat should be confined to this crate for two to three weeks. Though this might feel tough for both the cat and owner, it helps the cat acclimate to the barn and establishes it as home, reducing the risk of the cat wandering into dangerous territories.

During the training period, ensure the cat gets sufficient exercise. Once freed, continue to build a bond with your barn cat through consistent feeding routines, usually twice daily, even if they supplement their diet by hunting. This fosters a sense of security and reinforces the cat's connection to its home. A balance between affection and independence is key. Your barn cat is both a companion and a working animal. Cats that are well socialized can be incredibly friendly and effective hunters, offering the best of both worlds. Also, it is wildly irresponsible to have a barn cat that is still able to reproduce. It's your job to ensure that they are spayed or neutered.

It's essential to provide a safe, designated space within the barn that the cat can call its own. This area should be sheltered from the elements and offer protection from predators. A cozy bedding spot, clean water, and food are must-haves. Additionally, regular veterinary care ensures the cat remains healthy and capable of performing its duties.

However, it's essential to consider the broader ecological impact of introducing a barn cat into your farm environment. Cats are natural predators and, if not carefully managed, can threaten local wildlife, particularly birds and small mammals. To minimize ecological impact, ensure your cat's hunting instincts are primarily focused on rodent pests rather than native species. Providing adequate food and shelter can reduce the likelihood of overhunting, helping to maintain a balanced ecosystem while addressing your farm's needs.

A cat's murderous nature is one of the reasons I'm not a cat person.

By the fall of 2018, Lil Barn Cat had grown into her role on the farm. The tiny kitten who once trembled in the corner of the mudroom was now a sleek and confident young hunter. She'd graduated from supervised outings to roaming the farm on her own during the day, her striped tail held high as she prowled the fields and barns.

But becoming a true barn cat meant more than just day patrols. It meant being out at night, when the real work began. It wasn't a decision I made lightly. It wasn't a decision at all, really. It was just what barn cats do. Keeping her locked up forever wasn't an option, not if she was going to do her job.

One cool evening in early October, I crouched beside her in the mudroom, her tiny brass bell jingling faintly as she padded toward me. I reached down and unbuckled her collar, the one she'd worn as a padawan learner to help me keep track of her during her training months.

"All right, Lil," I said, holding the bell in my hand. "No more training wheels. You're officially a barn cat now. Go out there and make Pablo proud."

She looked up at me with wide, unblinking eyes, then rubbed against my leg as if to say, *Got it, boss.*

I watched as she sauntered out the door, her small frame disappearing into the dusk. Pablo waited for her by the barn, his larger, grizzled figure silhouetted against the fading light. He'd taken her under his wing in a way I hadn't expected, teaching her the rhythms of barn cat life. Together, they were a formidable team, grooming each other, hunting side by side, and patrolling the property like tiny sheriffs.

A few weeks later, I slept uneasily. I told myself it was silly. Pablo was out there with her, and she knew the terrain by heart. But around

3:00 a.m., a sharp, high-pitched shriek cut through the stillness, yanking me from sleep like an alarm.

I bolted upright, threw on my boots, and grabbed a flashlight. The house was cold, the kind of cold that makes everything sharper and meaner. I threw open the door, stepping into the icy night as the flashlight beam swept across the barn and fields.

"Lil!" My voice cracked, louder than I intended.

Pablo was sitting on the porch, his green eyes glowing eerily in the light. He looked stressed.

"Where is she?" I asked him, even though I knew he wouldn't answer.

The flashlight's beam swept across the yard, catching on shadows that looked far more sinister in the darkness. The barn, the chicken coop, the woodpile. Nothing. I called her name again, my voice cracking slightly.

By 4:00 a.m., I was pacing the yard, my breath visible in the cold air. My mind raced through every horrible possibility: coyotes, foxes, owls. I knew the risks of letting her live outside, and I thought I'd made peace with them. But now, standing in the dark with nothing but the wind and Pablo's unhelpful gaze, I felt the weight of it pressing down on me.

At first light, I gave up. The farm was stirring to life, the ducks beginning their morning chorus, and the sky softening from black to gray. My legs ached from hours of searching, and my throat felt raw from calling her name.

And then, just as I turned toward the house, I saw her.

Lil sat in the driveway, licking her paw as if nothing had happened. Her fur was slightly ruffled, but otherwise, she looked unharmed. She glanced up at me, her eyes wide and bright, as if to say, *What's the big deal?*

"Where have you been?" I asked, crouching down beside her. She rubbed against my hand, her small frame warm and solid, and let out a soft, reassuring meow.

I carried her inside, setting her gently in the mudroom. She stretched, her claws catching on the blanket we'd left out for her, and flopped onto her side with an air of utter nonchalance.

I sat on the floor beside her, my back against the wall, and let out a long breath.

"You're going to give me a heart attack, you know that?" I said.

She responded by batting lazily at the flashlight I'd set down, her movements unhurried but deliberate. I picked up a piece of straw and dangled it in front of her. She swiped at it once, twice, then grabbed it with both paws, holding it triumphantly like a trophy.

As we played, the tension in my chest began to ease. I'd spent the whole night worried sick over this tiny creature, but now, watching her bat the straw with the enthusiasm of a kitten, I couldn't help but feel relieved and, if I was honest, a little bit in awe.

I was still not a cat person.

By the summer of 2020, Lil Barn Cat had transformed from a scrappy rookie to a hardened pro. She was no longer just Pablo's sidekick. She'd earned her place on the farm as a pint-sized enforcer with a passion for murdering rodents. Sure, Pablo was still the elder statesman, but Lil had carved out her own identity. She wasn't just a barn cat. She was *my* barn cat.

It was a warm summer morning, and I'd just finished building a new chicken tractor, preparing it for a fresh batch of teenage goslings. The sun was rising high, the air buzzing with the soft hum of insects, and for a moment, everything felt calm.

Then, a car pulled into the driveway.

I wasn't expecting anyone, but it wasn't unusual for neighbors to stop by unannounced. I set down my tools and walked toward the vehicle, squinting against the glare. The driver rolled down his window, and his expression told me everything before he even spoke.

"I think I hit one of your cats," he said, his voice heavy with regret.

The words landed like a punch to the gut.

"Where?" I asked, my voice tight, already dreading the answer.

He gestured vaguely toward the road. "Out by the tree line," he said. "I didn't see anything there when I checked. I'm sorry."

I nodded, my throat too tight to respond, and he drove off, his car disappearing down the road.

"Allison!" I called toward the house, my voice sharper than I intended. She came running, her face pale.

"What happened?"

"Lil," I said, my chest tightening as the word left my mouth. "We need to look for her."

We grabbed flashlights even though it was broad daylight, driven by sheer panic. We scoured the property, calling her name until our voices echoed off the barn and into the trees. The barn, the fields, the hedgerows. We even checked the ditches along the road. Nothing.

By nightfall, I was walking the yard, my flashlight beam sweeping over the same areas I'd checked a dozen times already. My chest ached with worry and guilt. Lil wasn't just a cat anymore. She was family. And the thought of losing her was unbearable. I cried myself to sleep that night, grieving a cat that I never wanted.

That next day was hard. I recorded a YouTube video describing what had happened the day before. Looking back on that recording now, it was mostly tears and snot. I sleepwalked through the day, and by that night, I was ready to turn in early.

"Come downstairs! Quick! She's here," screamed Allison from downstairs.

I ran down to the porch, where Allison was crouched, her hands hovering near a small, hunched figure on the ground.

Lil.

She was sitting in the driveway, her fur matted and dull, her body trembling.

"She's alive," Allison whispered.

I reached out carefully, my hand shaking as I stroked the fur between her ears. "Hey, Lil," I murmured. "It's okay. We've got you."

Her tiny body was tense, but when I slid my hand under her belly to lift her, she didn't resist. I carried her inside, cradling her like she was made of glass. Allison grabbed the keys, and we drove straight to the emergency vet, the truck's engine humming loudly in the quiet night.

As I drove, my free hand rested gently on Lil's side, feeling the faint rise and fall of her breath. Allison reached over, placing her hand over mine.

"She's going to be okay," she said softly.

I nodded, though I wasn't sure if I believed it.

And I still wasn't a cat person.

The drive to Littleton, New Hampshire, was a blur. The emergency vet was just over the state line, the closest place we could take Lil in the middle of the night. I gripped the steering wheel tightly, my free hand resting gently on Lil's fragile frame as she lay on Allison's lap. Her breathing was shallow, her tiny body tense but still warm against my fingers.

The road ahead was dark and winding, and the truck's headlights cast long, jittery shadows on the trees lining the highway. Allison whispered to Lil the entire way, her voice low and soothing, as if her words alone could keep her alive.

When we arrived, the vet techs met us at the door. They moved quickly but carefully, lifting Lil from Allison's arms and whisking her into the back room. The antiseptic smell of the clinic hit me like a wall, and I stood there outside the clinic. Because of COVID protocols, we couldn't go inside.

"Let's sit," Allison said, touching my shoulder. Her face was pale, her eyes red from crying, but her voice was steady.

We sat in silence on a bench outside the clinic. Time seemed to stretch, every second feeling heavier than the last. Finally, the vet emerged, her expression serious but not hopeless.

"She's stable for now," she said, glancing at her clipboard. "But her injuries are significant. Her pelvis is shattered, and we're seeing signs of internal bleeding. We've done what we can to stabilize her, but she's going to need surgery. A specialist."

"What kind of specialist?" I asked, my voice strained.

"There's a veterinary hospital in Burlington," she said. "They have the equipment and expertise to handle this. But you'll need to take her there immediately."

Burlington was nearly two hours away. My stomach tightened at the thought of putting Lil through another car ride, but there was no choice.

"She's strong," the vet added, as if reading my mind. "She's made it this far."

We nodded, thanking her and paying the bill before they brought Lil out to us, bundled carefully in a soft blanket. I carried her to the truck, her small body lighter than I remembered.

"All right, Lil," I murmured as I settled her on Allison's lap. "One more ride. Hang in there."

"You know," Allison smiled. "I think this makes you a cat person."

"That's a dirty lie."

Twenty-four hours after finding Lil on the driveway, I was sitting in the parking lot of Burlington Emergency & Veterinary Specialists, staring at the windshield as a light rain drizzled down. The world outside felt muted, the gray skies mirroring the weight in my chest. COVID restrictions meant no one could go inside with their pets, so we'd handed Lil off to a vet tech at the door and now waited for updates by phone.

Allison sat beside me, her hand resting on my knee, her touch the only thing keeping me grounded. Lil was somewhere inside that building, fighting for her life.

When my phone finally rang, the name "BEVS" lighting up the screen, I grabbed it so fast I nearly dropped it.

"This is Dr. Snodgrass," the voice on the other end said. "We've reviewed Lil's condition, and it's critical. Her pelvis is shattered, her bladder has ruptured, and her blood is becoming toxic from internal complications. She needs emergency surgery to repair the bladder and stabilize her. Without it…" The sentence trailed off, but the implication hit hard.

"How much?" I asked, my voice tight.

"The first surgery will run between six thousand and eight thousand dollars," the vet said. "But she'll need a second procedure to reconstruct her pelvis, plus days of hospitalization. You're looking at a total of sixteen thousand dollars or more."

The number was staggering.

"Do it," I said without a hint of hesitation.

"Are you sure?" the vet asked. "It's a significant cost."

"Yes," I said, my voice steady despite the knot in my chest. "Whatever it takes."

I felt Allison squeeze my knee. When I glanced at her, she was crying, but there was a faint smile on her face.

"Thank you for letting us help her," the doctor said. "We'll get started right away."

The call ended, leaving us in a tense silence broken only by the rhythm of raindrops on the truck's roof. I leaned back in my seat, staring out at the rows of parked cars and wondering how many other people were sitting there, waiting for news like ours.

"She's strong," Allison said quietly, more to herself than to me.

I nodded, swallowing hard. "She has to be."

Those forty-eight hours felt endless. Each time the phone rang, my pulse raced, my thoughts spiraling between hope and dread.

"She's strong," Allison said again and again, like a mantra.

"She has to be," I replied, though my voice didn't carry the same conviction.

The clinic offered to let us FaceTime with Lil during her recovery. At first, I balked.

"Why would we FaceTime a cat?" I asked, my tone sharper than I intended.

"Because she needs to hear your voice," Allison said simply. "And maybe I need to hear it too."

So I did.

The first call was jarring. Seeing Lil hooked up to monitors and IVs, her small body barely visible beneath a nest of tubes and wires, felt like a punch to the gut. She looked so fragile, so impossibly small, but then her ear twitched. Just slightly.

"Hey, Lil," I murmured, my throat tight. "You're looking a bit rough, girl."

She didn't respond, of course, but I kept talking. I told her about the ducks and the geese, about how Pablo was still patrolling the barn like a grumpy old sheriff. Allison leaned into my shoulder as we watched Lil on the tiny screen, her chest rising and falling in slow, steady breaths.

"She knows you're there," Allison said softly.

I nodded, swallowing hard. "She'd better."

The calls became a daily ritual. Each time, I'd find something new to tell her, filling the silences with updates about the farm or small reassurances that she was doing great. It felt ridiculous at first, but over time, those moments became something I looked forward to. It was a way to remind myself, and maybe her, that we weren't giving up.

By the end of the week, the vets determined she was ready for the second surgery. This time, they focused on her pelvis, using screws and plates to piece together the damage. The procedure was risky, but it was her best shot at regaining mobility.

"She made it," the surgeon told us afterward, his voice tinged with exhaustion and triumph. "Her recovery will be long and difficult, but she's got a chance."

I exhaled, the tension in my chest finally loosening.

"But I'm still not a cat person."

When the vets finally discharged Lil, they handed us a care plan that looked less like instructions and more like the script for a medical drama. Medications multiple times a day, movement restrictions so strict they bordered on ridiculous, and a dire warning: The next three months would decide everything.

"She'll need to stay in a crate," the vet said, their tone as clinical as the fluorescent lighting. "No jumping, no running, no climbing."

I nodded, but the words barely registered. This was Lil Barn Cat, *my* barn cat, the pint-sized enforcer who stalked fields and scaled hay bales with the confidence of someone who didn't know how small they were. And now? She was supposed to live in a crate for the next three months on movement restriction so that her broken body could heal.

The drive home was nerve-wracking, every bump in the road making me flinch. Lil lay limp in Allison's lap, her body unnervingly still. She was alive, but barely. The spark that had always defined her seemed dimmed, like a fire running low on fuel.

"You're home now," I said softly as we set her crate in the mudroom, the same place where her life on the farm had begun. "We've got you."

Lil blinked up at me, her green eyes heavy with exhaustion but still holding a flicker of that signature defiance.

The first few days were chaos. I spent more time crouched by the crate than doing anything else, coaxing her to take her meds as she swatted weakly but determinedly at my hands.

"Just hold her," Allison coached, her voice calm but firm. "She'll fight, but it's for her own good."

And fight she did. Every pill, every syringe of medication was a battle. She'd twist and yowl, her claws catching my skin as she reminded me, *I'm still me, don't you forget it.*

"You're doing good, Lillian," I whispered after one particularly ugly round. She blinked at me, her paw twitching like she wanted to swat but didn't have the energy.

Recovery was slow and grueling. The screw and plates in her pelvis made every movement a struggle. Even now, I can feel the faint bump of metal under her fur, a permanent reminder of what she went through. I sat beside her crate for hours, dangling pieces of straw or whispering about the farm, desperate for any sign of the playful, stubborn cat I knew.

But the hardest part wasn't the medical care. It was accepting what came next. Lil, the fierce barn prowler, was now an indoor cat. Temporarily, maybe, but the allergies that turned me into a sneezing, eye-rubbing wreck made the situation feel unsustainable.

"This can't work," I muttered one night as I rubbed my burning eyes for the hundredth time.

Allison smiled knowingly. "And yet, here you are."

She wasn't wrong. The mudroom wasn't just Lil's space now. It was the heart of the house, and moving her felt unthinkable. Lil wasn't just a barn cat anymore. She was *our* cat.

Then there was the matter of the cost. Two surgeries, weeks in the hospital, constant vet care. When the bills were finally tallied, the number hit like a punch to the gut. Even though we'd been warned.

I couldn't stop thinking about everything that money could've gone toward—a farm vehicle, new equipment, anything. The guilt gnawed at me, but so did the knowledge that it had been worth it.

Asking for help has never been something I've felt comfortable with. I'd been that way since I was a kid, stubbornly determined to handle things independently, even when it wasn't the best choice. It's a toxic trait, sure, but it was one I'd carried with me well into adulthood.

After posting a video about what had happened to Lil, I thought I was just sharing her story, trying to process everything. I never imagined what would happen next. At first, it was just a few small things. A check in the mail, a note from someone who had been watching the channel, offering to help. Then the PayPal deposits started coming in. One after another, from people I'd never met, who'd heard about Lil and just wanted to do something.

It felt surreal, and honestly, it made me deeply uncomfortable. I wasn't used to accepting help, especially not like this. I kept thinking about all the people who were struggling so much more than we were, who needed that money more than I did. It didn't matter that the vet bills were astronomical, or that every dollar was going toward saving Lil. I couldn't shake the awkwardness of it. All told, Gold Shaw Farm viewers ended up paying for roughly a third of the vet bills.

Looking back, I'm still floored by the generosity we received during that time. Every check, every PayPal deposit, every order for a T-shirt felt like a small miracle. It reminded me of the scene in *It's a Wonderful*

Life when George Bailey realizes the whole town has shown up to help him. It's a moment that always makes me tear up, and even now, when I think about what people did for us, I feel the same way.

That experience shaped me. It's part of why I try so hard to pay it forward now, through the charitable work we do on the farm and the fundraising efforts we've taken on. It taught me that sometimes, people just want to help. And accepting that help doesn't make you weak. It makes you part of something bigger.

Even today, I feel awkward about the generosity we received, but I'll never stop being grateful for it. And I've tried to do my best to pay that kindness forward.

But that still doesn't make me a cat person.

These days, Lil Barn Cat isn't prowling the fields or staking her claim on the barn rafters. No, she's traded her feral past for a life of indoor opulence, ruling over my office like a queen surveying her kingdom. Her favorite perch is the top of the cat tree I bought her—a throne, really—strategically placed by the window for maximum yard-monitoring efficiency. When she's not there, she's stretched out by the fireplace or sunbathing in a warm patch of light. And wherever I am, she's not far behind.

Turning Lil from a scrappy, wild barn cat into a house cat wasn't the plan. It was the opposite of the plan. I'd always believed barn cats belonged outdoors, free to roam and keep the rodent population in check. Bringing her inside felt like betraying the very concept of what a barn cat was supposed to be and a direct affront to my allergy-prone existence. But, like Lil herself, life has a way of ignoring rules.

At first, it was a necessity. Her injuries made the barn impossible, and the mudroom became her convalescent home. I thought it was temporary. She had other ideas. Lil adjusted to indoor life with unnerving speed, as though she'd been waiting for central heating her whole life.

Not that it was easy. Those first few months were brutal: sneezing fits, watery eyes, Benadryl by the fistful. There were nights when I lay awake, sniffling and wondering if I'd made a colossal mistake. But slowly, something changed.

Maybe it's the same logic as allergy shots. Exposing yourself to tiny doses of what you're allergic to until your body stops freaking out. Or maybe it's just that I had no choice but to adapt. Either way, my sneezing became less frequent, my eyes stopped burning, and I gradually phased out the nightly Benadryl routine. I'm no doctor, but my theory is simple: Lil made me tougher in more ways than one.

Despite the hurdles, Lil became my constant companion. She's always nearby, watching as I edit videos, ignore emails, or draft a new book. In fact, she's been here with me for every single word I've written for you in this book.

Sometimes she naps. Sometimes, she yells at me for food. And sometimes, when I need to think about something, I'll scratch that soft spot just behind her ears and murmur to no one in particular, "I'm still not a cat person."

CHAPTER 10

WILL OWNING CATTLE MAKE ME A REAL FARMER?

Toby Dog stood motionless beside me, ears perked, nose twitching at the crisp October air. He could smell my anxiety. The pasture stretched before us, golden in the early autumn sun, still damp from morning dew. The pasture was empty. Just an expanse of open grass and taut electric polywire strung with care and a certain level of panic. My heart was thumping, a rhythmic blend of excitement and dread. It felt like standing backstage moments before walking out in front of an expectant audience.

"What do you think, Tobes? Ready to be a cow dog?" He tilted his head, the universal canine expression for *you're talking nonsense again, human.*

But that was just it. I didn't know if I was ready. Ducks, chickens, and geese were manageable. Small, quirky creatures I could easily scoop up with a fishing net. But cattle? Massive, powerful animals that could trample fences, break bones, and wreak genuine havoc across the landscape. The decision felt heavy, as if buying cattle would finally grant me entry into the fraternity of farmers who earned respect through sweat, scars, and stubborn endurance. It was the thing that would prove I was more than just some hobbyist or internet personality playing farmer. Cattle would make me a real farmer.

Mob grazing, the ancient-but-now-new practice of regularly moving cattle to mimic the natural patterns of wild ruminant herds, had long captivated me. The image of cattle tightly grouped, consuming grass, trampling weeds, and fertilizing the soil in a choreographed dance had inspired me since I first read about the practice. I wanted that here on my farm. That ancient dance between hoof and soil. The biting of grass. The dropping of poop.

Toby Dog smacked my leg with his forepaw, snapping me back to the present.

"Yeah, I know," I muttered, rubbing his head. "We'll manage. We always do."

Just then, I heard a moo. I turned and waved as my neighbor Donnie's old Ford rolled into view, the livestock trailer bouncing over the uneven farm road that ran alongside the barn. Donnie had lived in Peacham all his life. A seasoned cattleman who wore his weathered chore coat like a uniform. He was helping me transport my herd from a farm out in South Hero, Vermont.

Donnie pulled alongside me and killed the engine, leaning out the open window. "You sure you're ready for this?"

"Ready as I'll ever be," I lied, plastering on a confident grin.

He chuckled, not unkindly. "Well, just so you know, these cattle ain't exactly lap dogs. Better keep your eyes open and avoid ending up on the business end of their horns."

Ugh. Why did I have to get the shaggy cattle with the three-foot-long horns jutting out of their skulls?

"Yeah, the guy who sold them to me mentioned something about getting gored recently. Very reassuring, right?"

"Yep," Donnie laughed dryly, his eyes twinkling with mischief. "Welcome to the big leagues, kid."

Behind me, Donnie had hopped out, opening the rear trailer doors with a metallic creak that echoed through the pasture. He peered inside cautiously. I glanced from the trailer to the open field, feeling the weight of what was about to happen. My hands trembled slightly, adrenaline sharpening the world to the 4K high definition I could see through my camera's viewfinder.

Toby Dog barked.

"Yeah, buddy," I muttered, half-smiling, half-wincing. "This could go sideways real quick."

A deep breath. A steadying nod.

"Ready?" Donnie shouted.

I turned the camera toward the trailer, squared my shoulders, and faced my audience—both the ones in front of me now and the ones who would watch this later, dissecting every moment.

"I'm about to ruin my life," I said, grinning like I meant it. "Let's film this."

With a metallic clang, the trailer doors swung open.

ChoadToasty421: He's a hobbyist. All he's got is a bunch of ducks.

Ever since I was a kid, I've always been chasing something. But at various points in my life, I haven't been quite sure about the *what.* A feeling, maybe. An identity. A version of myself that hovered just out of reach, shifting every time I thought I had it pinned down. Ever since moving to Vermont, buying a farm, and throwing my entire life online, the same question had haunted me: Was I just a live-action role-player, an entertainment spectacle for the masses?

pungus_france: Q: What's the difference between a farmer and LARPer? A: LARPers have more subscribers than animals.

I started small, cautiously, dipping my toes into the farming world with ducks, chickens, and geese. Manageable creatures. Lightweight. Easy to scoop up and relocate without heavy machinery or existential dread. They were my training wheels, my gateway livestock. But deep down, I knew the truth. Ducks and geese didn't quiet the voice in my head that whispered: *Real farmers have cattle.*

It wasn't just in my head, either. The comments section under every YouTube video was a relentless chorus.

CluckNorrisOfficial: He's not a farmer. just another city guy dressing up.

AlfalfaMale: Oh look, he thinks he's a farmer. Cute.

IH8Barns69: Let us know when your graduated to real livestock.

Each comment hurt more than I could admit. Was I just fooling myself?

One night, after a particularly biting string of comments, I found myself pacing the kitchen, staring at my phone screen. Allison sat at the table, thumbing through some book, half-listening.

"You know what someone said today? They said I was just 'playing farm.' Like it's some elaborate cosplay."

Allison looked up, eyebrows raised. "Do you believe them?"

"No," I said too quickly, but then my tone softened. "Maybe?"

She closed her book, sliding it across the coffee table with a sigh. "Who cares what some rando says?"

"But it's not just random trolls. It feels real. Like there's an invisible line between real farmers and everyone else. And I'm always on the wrong side of it."

She smiled gently, leaning back in her chair. "What's gonna make you feel like you've crossed that line? What's your magical milestone?"

Ever since I had contracted the agriculture bug back in DC, I had been obsessed with the idea of a cattle farm. I pored over books and videos from regenerative farmers like Joel Salatin and Greg Judy, trying to grasp the intricacies of mob grazing. This nature-mimicking practice resonated with something deeper inside me. Mob grazing was not merely farming. It was orchestrating the dance of nature itself.

MilkinIt4Clout: This guy is all fat and no cattle.

As for choosing a breed? That part was less profound. No grand vision of genetic superiority, no master plan. Just a practical checklist: small enough to manage, hardy enough for Vermont winters, willing to eat marginal forage. Enter: Scottish Highland cattle. Because nothing says "like and subscribe" quite like a cow that looks like it fronts a jam band.

Scottish Highland cattle look like something out of a medieval tapestry: shaggy, russet-coated beasts with sweeping, curved horns that make them appear equal parts regal and prehistoric. Bred for the harsh landscapes of the Scottish Highlands, they evolved to thrive in brutal winters, their thick double coats shielding them from snow and wind, eliminating the need for excessive fat stores, making their beef leaner than many other cattle breeds. Beneath their woolly exteriors, Highlands are hardy, self-sufficient grazers. They are capable of eating marginal forage that other cattle might struggle with. Their lineage stretches back centuries, traced to the ancient Celtic cattle that roamed Scotland, shaped more by survival than human intervention. Highlands are famously docile, intelligent, and adaptable despite their imposing appearance. Gentle giants with an ironclad ability to turn rugged land into thriving pastures.

ChicknTendieTruthr: Of course he's getting highlanders. He probably bought them at the poster section of Hobby Lobby.

When I finally found a farmer in South Hero selling Scottish Highland cattle, excitement rushed over me in waves. Picking out my herd felt ceremonial, an initiation into something bigger than myself. Five thousand dollars later, I was now the proud owner of one shaggy steer and four pregnant cows. Standing there, bill of sale in hand, it felt like receiving a secret pass into a world that might never fully accept me.

But when the cattle arrived, big, anxious, wild-eyed, I felt something entirely unexpected. It wasn't validation or triumph. It was responsibility. It was the sudden weight of more lives depending on me. That day, standing by the trailer, watching them nervously step onto my land, I'd expected to feel transformed. Instead, I felt like a fraud who had simply raised the stakes while holding a crap hand.

What had I done? My heart rate instantly doubled. Five cattle. Half a ton each. Vanished in less time than it took to make a sandwich. I stood at the gate, fingers still gripping the cold metal, adrenaline surging through my veins. Toby Dog barked urgently, his tail stiff, ears swiveling toward the orchard. He sensed it, too. Chaos had been unleashed at the Gold Shaw Farm.

"Dammit," I muttered. "Where did they go?"

Less than an hour ago, everything had been fine. I'd watched them graze peacefully behind the strands of electric polywire, feeling smug about how smoothly they'd settled in. I'd even allowed myself a moment of self-congratulation before stepping inside to inhale a sandwich. Rookie mistake.

But when I'd come back out, the cattle had vanished like a five-thousand-dollar magic trick gone terribly wrong.

I sprinted uphill, boots slipping on slick October grass. Toby Dog bolted ahead, barking wildly, his nose locked onto the scent of trouble. And there they were. Five shaggy escape artists ambled through my orchard, stomping over roots I had carefully cultivated for years. Annabelle had a branch in her mouth, chewing methodically. Anne of Green Gables swung her horns dangerously close to a chestnut sapling, utterly unbothered by my rising panic. Kurt Cobain, my scruffy young steer, was busy *pawing* at the roots of a four-year-old tree like he was auditioning for a demolition crew.

"Hey!" I shouted, waving my arms. "Get away from there!"

Ariel lifted her head sharply, eyes locking onto mine. She didn't bolt, but she didn't back down either. Toby Dog growled low, his body tense. I grabbed his collar, whispering, "Easy, buddy. Don't spook 'em."

How had this unraveled so fast?

In my excitement, I'd made a critical mistake. Two measly strands of polywire. Cheap, easy, and laughably inadequate for cattle who'd never felt the bite of electricity before. From my vantage point, I could see the snapped fencing strands waving in the breeze, as useless as a wet noodle. They had probably brushed through it without even noticing.

"Toby Dog, we gotta fix this. Now."

I turned and sprinted for the ATV, grabbing reels of thicker, stronger polyrope, my breath ragged, hands already shaking from the adrenaline dump. Every second counted. This was real farming, where fences failed, animals panicked, and every mistake carried a price tag. The kind of thing YouTube edits out.

As I worked, I mentally tallied the herd.

Four.

I counted again.

Four.

"Where did that Anne Shirley girl go?" I Marilla Cuthbert-ed.

My mind raced: Loose on the road? Tangled in ancient barbed wire? Trampling a neighbor's garden? How many ways could this go horribly wrong?

Toby's ears perked, head snapping toward the south end of the orchard. He barked sharply.

"Move carefully, move quietly," I reminded myself, unspooling the electrified rope.

They ignored me, leisurely chewing and ambling further into the orchard. My blood boiled with frustration. This wasn't ducks escaping through a hole in the poultry netting. These were half-ton animals, each capable of snapping bone or destroying trees with a careless flick of its head. The weight of what I'd taken on suddenly crashed down on me.

It took three exhausting, muscle-burning, sanity-testing days before I properly corralled them. Each night, I returned exhausted, sweaty, and sore from fencing work. Yet, each morning, I got up and did it again. I had to.

Those early weeks felt like boot camp. And despite small victories, I never quite shook the creeping fear. I imagined seasoned farmers smirking knowingly as they watched my videos, waiting patiently for me to fail.

CropDuster1337: Of course, this guy would lose his cattle in the first five minutes.

SquirrelyDanStan: He's walking around pouting like his little sister just ate his last Oreo!

Late one evening, while rewatching raw footage from the previous day's video, I asked aloud the question that haunted me: "What if they're right? What if I can't handle this?"

The only reply came from Lil Barn Cat, who yowled a demand for more cat food.

Three years of cattle farming came and went in the blink of an eye. Although I always had perimeter fences in place, my cattle escaped dozens of times over the years. The experiences had left me humbled, bruised, and vaguely feral. But I had also evolved. I now wielded an industrial-strength electric fence charger capable of frying the soul of a careless human or a curious livestock guardian puppy. I could move an entire herd using only my voice and a handful of alfalfa cubes, essentially wizardry. I had mastered the delicate ballet of calving season, veterinary schedules, and the endless, soul-sucking grind of cleaning out semi-frozen cow poops out of winter paddocks.

My cattle had even earned names, an elaborate, silly-but-systematic naming tradition. Every first-generation female carried a name beginning with *A*. Annabel, Audrey I, Ariel the Little Mermaid, Anne of Green Gables, Amanda Hugginkiss, Astrid, Amelia Bedelia, Alice B. Toklas. Their offspring moved down the alphabet, a new generation carrying forward a legacy I'd painstakingly crafted. Belinda Carlisle, Bonnie McMurray, Beatrix Potter, Belle, Betty Rubble, Betty White, Buffy the Vampire Slayer. Steers got immortalized differently, named after dead rock stars. A gentle reminder of their purpose and short life expectancy. Jimi Hendrix, Joey Ramone, Kurt Cobain. Naming them made it personal. Too personal maybe, but I think the personal connection to the animal is crucial for their entire time on the farm. Even if they are ultimately destined for the dinner plate. Giving them a name is a sign of respect.

And, of course, there was Macho Man Randy Savage, the charismatic cream of the crop at the center of it all. Macho Man wasn't just a bull. He was *the* bull. And, if I'm being honest, the idea of owning a bull initially terrified me.

Most of my cattle are of unregistered origins. I figured it didn't matter too much since they were beef animals. But since the father's genetics are so important and I was very nervous about buying a bull who was as wild as my cattle initially had been, I decided to splurge a bit.

Bulls have a reputation. They're massive, unpredictable, and dangerous. More than one farmer has a horror story about a bull that turned mean, and I didn't particularly want to become one of those stories. But Ray and Janet of Shat Acres Farm in Plainfield, Vermont, the couple who bred and raised Macho Man, set me on the right path. Their farm was the kind of place that made you rethink what cattle could be. Every animal was calm, socialized, and handled daily. Proof that a bull didn't have to be a walking liability. Under their mentorship, and with a whole lot of time spent brushing, halter training, and just being around him, Macho Man slowly shifted from an intimidating presence to my favorite bovine on the farm. He wasn't just tolerable. He was enjoyable. I could lead him on a halter. I could scratch his neck. I could work alongside him without that gnawing pit of anxiety in my stomach. And somewhere in that process, my fear turned into trust. I still respected his size and his power because that's the first rule of working with bulls. But I didn't see him as a threat anymore. I saw him as a partner. And for a guy who once swore he'd never own a bull, that felt like a pretty big deal.

Those first Vermont winters hit me like an icy fist. Vermont winters were harsh. Hay bales stacked like gold bricks disappeared alarmingly fast. Vet bills piled up, infrastructure needs multiplied, and my bank account shrank with frightening efficiency. I learned to wake early, moving frozen bales of hay, breaking ice off water troughs, slipping in manure and getting right back up again, day after relentless day. I developed calluses, bruises, and a constant state of fatigue that coffee could never fully erase. Still, I pressed on, stubbornly optimistic, committed to proving myself.

But then came the spring of 2024.

Six calves were born that season, each birth an exciting farm memory. Until two calves died within a gut-wrenching span of just ten days. Elliott Smith and Prince, two healthy-looking bull calves, succumbed mysteriously. I still get sad when I think about how I held baby Elliott in my arms for hours, desperately trying to get him to stand, feeding him by hand, until the light left his eyes late one night. I buried both calves myself, digging into recently thawed soil by hand.

When the necropsy lab results landed in my inbox, guilt slammed me like a kick in the knockers. It was a selenium deficiency, such a basic mistake. My fault.

Lactose_Intolerant860: Bro pretended so much he killed them calfs.

The shame burned deep, confirming my worst fears: Maybe I really was just pretending. Would this ever stop hurting? Could I live with these mistakes? Was I truly a farmer, or had I just raised the stakes in an elaborate, public game that I wasn't cut out to win? Poultry was easier and far less heartbreaking. It was right around that time that I seriously considered quitting farming, filming, and filming farming.

PART THREE

EMBRACING THE MESS

CHAPTER 11

PAYING PEOPLE TO BUILD IKEA FURNITURE

I am losing a battle against a chair.

Not just any chair. A sixty-nine-dollar IKEA chair with a name that sounds like an unpronounceable Swedish venereal disease. Grönsta. A chair that exists only in theory currently manifesting as a jumble of particleboard, screws the size of baby teeth, and a single, condescending Allen wrench.

The Allen wrench sits heavy in my palm, a tiny instrument of doom. I glare at the instruction booklet, a cryptic series of pictograms that may as well be hieroglyphics. The little cartoon man is there, as always, smiling blankly as if to say, *You are not smart enough for this task.*

"You smug little cartoon bastard," I mutter.

The assembly process is allegedly simple: align the holes, insert the dowel, tighten the screw. But nothing aligns. Nothing inserts. The pieces resist each other like repelling magnets, as if this chair does not *want* to be a chair.

How had it come to this?

I came from sturdy stock. Generations of practical, self-sufficient men who built things with their own hands. My grandfather had been a plumber. Most of my cousins and uncles were electricians. But me? I was a thirty-three-year-old man incapable of putting together a sixty-nine-dollar chair without spiraling into an existential crisis.

I try Google. It offers only blurry YouTube tutorials, an assortment of dad-blog message boards, and, unhelpfully, ads for other chairs. My forehead presses against the unfinished seat. IKEA is winning. I am losing.

Eventually, I cave. I open TaskRabbit, tap a few buttons, and, like a miracle summoned from the heavens, a man called Darnell appears at my door with a toolbox and a calm demeanor that suggests he has never once considered throwing a half-built bookshelf off a balcony.

He assembles the chair in under twenty minutes. I do not speak. I just watch, hollow-eyed, as he tightens the last screw with the ease of a man tying his shoes.

When he leaves, I sit down. My new chair is sturdy, well-built, and profoundly disappointing.

Maybe this is just how life works now. Some people build things. Others pay people to build things for them.

That's fine.

Right?

Two years after my bout of Grönsta, I found myself in the aisles of a Home Depot in Washington, DC, holding a power drill like it was an alien artifact. The move to DC had been meant as a reset, a new city, new job, new habits, new me. And part of that "new me" plan involved a noble, romantic notion of self-sufficiency. The kind of person who looked at a pile of lumber and saw potential instead of a future trip to urgent care.

As discussed in an earlier chapter of this book, I had been seduced by farmers' markets. The dirt-dusted carrots, the people in linen shirts earnestly discussing the terroir of their tomatoes all felt so *real,* so grounded. Unlike IKEA furniture, these things had weight, imperfections, history. Which is how I convinced myself that I, a man who once outsourced a chair, was going to build my own raised garden beds so I could grow my own food.

It seemed reasonable. Garden beds were just rectangles. Four boards, some screws, a little determination, essentially a box without a top or bottom. How hard could it be?

When I mentioned I was going to build the garden beds myself, Allison looked up from her tea and said, "You mean like…with wood?"

I nodded.

"And tools?"

I nodded again, slower this time.

There was a long pause.

"Huh," she said, in the same tone you might use when a dog opens a door by itself. Not disbelief exactly, but more like cautious curiosity about what might happen next.

"That's the plan," I said, puffing myself up a little. "How hard could it be?"

She didn't answer. Just sipped her tea and gave me a look I would come to recognize often in the coming years: the early warning signal for when she'd decided to let me learn something the hard way.

I had a list:

- Wood (the right kind? Unclear)
- Screws (size unknown)
- A drill (seemed essential)
- A saw (terrifying but also, apparently, necessary)

A helpful employee approached. *Did I need assistance?*

Yes. Deeply. On the most fundamental level possible.

"Nope, all good," I said and proceeded to buy everything blindly, as if I were on autopilot in a grocery store and not about to take possession of *a saw*.

Back home, I laid out my supplies in the backyard. The wood stared at me. I stared back. We were at an impasse.

YouTube, my ever-patient mentor, came through. I found a video hosted by "Lawn Order: Special Mulching Unit," featuring a man whose calm, confident voice suggested he'd never once doubted his gardening prowess. He made it look effortless. He was a friggin' liar.

I spent the next four hours sweating, swearing, and nearly impaling my own thigh with a drill. But somehow, I did it. Four walls that could hold dirt. Screwed together. A rectangle.

I stepped back and examined my creation. It was…*fine*. The corners didn't quite meet. One board had a slight warp, as if protesting its new existence. Nothing was square, but I didn't even know what that term even meant at the time.

But it was *mine*.

For the first time in my life, I had built something with my hands. I ran my fingers over the rough wood, feeling its weight, its imperfections, the undeniable fact of its existence. No TaskRabbit. No hired savior. Just me, my own questionable craftsmanship, and a deeply unsafe use of power tools.

I was not about to start building furniture from scratch. But for the first time, self-sufficiency didn't feel like a complete joke.

Of course, that illusion lasted *exactly* three weeks.

Because when my next IKEA purchase arrived, I did what any reasonable person would do: I paid someone else to assemble it.

One garden bed did not make me a carpenter.

But at least I wasn't losing to chairs anymore.

By the time I moved to Vermont, I had convinced myself that I was ready for the rugged, self-sufficient life. I had built a single garden bed, after all. With my own two hands. Using tools. That had to count for something.

Vermont was not Washington, DC. It was not a place where people dabbled in rusticity on weekends before retreating to apartments where the water always worked and no one had to think about things like septic tanks or snow chains. It was a place where people built their own homes, fixed their own trucks, and probably didn't even buy IKEA furniture because they could just make their stupid chairs themselves out of the stupid trees they cut down themselves.

I was, in every possible way, out of my depth.

The first time I tried to fix a broken fence, I spent an hour staring at it, waiting for inspiration to strike. It did not. The first time I attempted to split firewood, I nearly took off my own foot. The first time I asked a neighbor if they knew a good handyman, they gave me a look that suggested I had just insulted their entire bloodline.

It was clear: I needed help.

And that's right around the time I met Alfred.

We were both waiting for our respective egg sandwiches at the Peacham Café, a tiny spot that doubled as the town's unofficial gathering place. I had started stopping in regularly, trying to get a feel for the community, mostly by sitting quietly and nodding at people like a friendly, socially awkward ghost.

Now if you've followed my YouTube channel for any length of time, you already know about My Buddy Alfred™. He is, in many ways, a mythical figure. A ninth-generation Vermonter. A stonemason by trade, but also a machinist, a fabricator, an inventor, a tailor, and an engineer in the way that only self-taught geniuses can be. If there is a problem anywhere in Peacham, be it mechanical, structural, or something as absurd as "How do we bend metal poles to build a greenhouse?" Alfred will find a way to fix it.

He is, in short, handy as hell.

Longtime video viewers will often get upset with me if I mention Alfred's name without playing his banjo-fueled theme song. (Yeah, a few years back, I commissioned a musician to record one.)

My buddy Alfred,
He lives on down the road,
He's a wizard with machines,
A genius in blue jeans,
Let's stop and say hello,
My buddy Alfred,
Vermont's best so-and-so,

Brick-laying engineer,
Surgeon driving heavy gear,
Ain't nothin' he don't know.

It's ridiculous, catchy, and it's entirely accurate.

But before Alfred became a recurring legend in my life, and on my YouTube channel, he was just a guy I met while waiting for an egg sandwich.

That morning, we started talking. I don't remember exactly how, but within minutes, Alfred told me about the maple syrup evaporator he had built by hand, out of bricks. A homemade maple syrup evaporator. I didn't even know what that was, and I was already deeply impressed.

Sensing my fascination (or possibly my incompetence), he invited me to come by his place and see it in action. I accepted immediately, partly because I wanted to see the evaporator but mainly because Alfred clearly possessed every skill I lacked. I needed to understand how a person like that functioned.

The next day, I followed a hand-drawn map to his house because GPS in Peacham was more of a suggestion than a reliable guide. The roads wound through hills and dense forests, and at least twice, I had to stop and convince myself that I wasn't about to drive straight into the plot of a horror movie. It was only about a mile and a half down the road from the farm.

When I finally arrived, Alfred was already outside, waiting. The evaporator was exactly what he had described: a functional, well-crafted, completely DIY setup that turned tree sap into liquid gold.

Of course, it was May, so maple sugaring season had just ended. But even without the boiling sap, I could tell: This was a dude who knew how to build things.

I stood there, taking it all in, feeling equal parts admiration and shame. Here was a guy who could construct an entire working system from scratch. Meanwhile, I had once declared victory over a garden bed that looked like a child's drawing of a box.

Alfred, of course, was unfazed by my awe. To him, this was just normal life.

"You ever tap trees before?" he asked.

I had not.

"Well," he said, gesturing toward the woods, "you'll need to learn for next season."

And that's how it started.

Over the next few months, Alfred became my unofficial mentor in all things practical. He didn't say so outright. There was no formal agreement, no syllabus, no "Self-Sufficiency for Flatlanders 101." But anytime I got myself into trouble, Alfred had a way of appearing, offering guidance in his dry, matter-of-fact way.

For example:

The time I tried to fix a broken gate by brute force, I nearly dislocated my shoulder.

"Did you check if the hinges were just rusted?" Alfred asked, barely suppressing a smirk as I stood there, panting, the gate as immovable as ever.

I had not checked. Turns out, the solution was not sheer force but WD-40.

One January morning, I tried to turn on the yard hydrant to get water to the ducks, and nothing came out. Naturally, I assumed the pipes were just haunted.

Alfred, when consulted, had a different theory.

"Your pipes are probably frozen," he said.

"What do I do?" I asked, expecting some kind of complex, homeowner-101 solution.

He shrugged. "Hair dryer."

This seemed like a joke, but it was not. Thirty minutes of blasting hot air from my wife's Conair, the water came back. Magic. Or, as Alfred would put it, *basic physics*.

It became clear that there was an entire *layer* of knowledge that I lacked, not just about *how* to do things, but *how to think about doing them*.

Alfred, for instance, never seemed *surprised* by a problem. Something was broken? Of course it was broken. Things break. You fix them. The idea that you would throw up your hands and panic simply did not compute in his brain.

Meanwhile, I had spent most of my adult life outsourcing my problems to other people. My dishwasher was making a weird noise? Call a repair guy. My car needed something beyond an oil change? Take it to a mechanic. My IKEA chair refused to be a chair? Pay some app-based startup to deal with it.

But in the Northeast Kingdom, that mindset didn't work. There were no on-demand services. No TaskRabbit, no Uber, no quick solutions. If something broke, you either figured out how to fix it, or you *waited a very long time* for someone who could.

And so, slowly, under Alfred's unspoken tutelage, I started learning.

Not in a graceful, Hollywood-montage kind of way. More in a "do it wrong, listen to Alfred explain what I should have done, and then do it slightly less wrong next time" kind of way.

It was humbling.

And yet, for all the frustration, minor injuries, and botched attempts, I found myself starting to enjoy it. There was something deeply satisfying about learning how to do things I had never even considered trying before.

One day, after successfully repairing a fence (without Alfred's direct intervention), I sent him a photo of my handiwork, feeling *just a little* proud of myself.

A few minutes later, he texted back:

"Looks good. It's backwards, though."

Progress.

It was July, and the ground was wetter than it should have been. A few months of relentless rain had turned parts of my pasture into something closer to a swamp, but I had convinced myself it wasn't *that* bad.

I had been out brush hogging, clearing tall grass, feeling good about tackling a necessary farm chore—one of those real, practical tasks that, for once, I actually knew how to do. And then, in an instant, I didn't feel good at all.

The back wheels of the tractor sank.

I tried to back up. The wheels sank deeper.

I put it in low gear, hoping to creep out of the muck. Nothing.

I tried a stick under the tires for traction. That only seemed to make the ground angrier. I had been on the farm long enough to know the truth: I was stuck. *Properly* stuck.

There was a time, not long before this, when I might have just panicked. Maybe Googled "how to get a tractor unstuck" and gone down a rabbit hole of forum arguments. Maybe spent hours trying increasingly bad solutions before caving and calling a tow service that would take two days to arrive.

But this wasn't Washington, DC. There were no on-demand services for stranded idiots in muddy fields.

So I pulled out my phone and texted Alfred.

Me: "Got the tractor stuck in the back pasture. It's bad."

Alfred: "I'll grab mine. Be there soon."

That was it. No lecture. No questions. Just a solution in motion.

Twenty minutes later, Alfred rumbled up in his own tractor, surveying the situation with an expression that hovered somewhere between amused and unimpressed.

"Picked a hell of a spot," he said, hopping down.

"Didn't *pick* it," I said. "It picked me."

He chuckled and got to work like he had done this a thousand times—which, knowing Alfred, he probably had. Chains. Tow straps. A little finesse with the throttle. Within minutes, he had my tractor pulled free like it had never been stuck in the first place.

I stood there, muddy and vaguely ashamed, expecting some kind of lesson. Alfred just dusted off his hands. "Happens to everybody. William had to pull me out just last week."

That was it.

No ridicule. No, *You should have known better.* Just a matter-of-fact acknowledgment that sometimes, no matter how competent you are, you're going to get stuck. And when you do, you need to know when to ask for help.

Somewhere along the way, I had absorbed the idea that a person either *knew* how to do things—build a fence, fix a truck, tap a maple tree—or they didn't. There was no middle ground. No learning curve. You were either handy, or you weren't. Competent, or clueless. An Alfred, or a Morgan.

But that wasn't how Alfred saw it. To him, everything was just a skill to be learned. You didn't have to be born knowing how to repair a tractor or install a water line. You just had to be *willing* to learn.

That was the crucial insight.

There was a time in my life when I might have seen this as a failure. A confirmation that I wasn't cut out for farm life, that I would always be an outsider. But standing there in the mud, watching Alfred coil up the chains, I realized I didn't feel that way at all.

Self-sufficiency wasn't about doing everything alone. It was about knowing how to *leverage* the resources around you, including the people who knew more than you did.

I was still learning. Still getting stuck. But at least now, I knew who to text.

I'm never going to be as handy as Alfred. I'll never suddenly develop an intuitive understanding of mechanics or wake up one morning with the ability to build a stone wall from scratch. But I had learned something just as important: I didn't need to.

I've long since stopped seeing competence as an all-or-nothing quality. I don't have to be a master of every skill to be capable. I just have to be willing to learn, to try, and, when necessary, to ask for help.

That realization changed the way I approach everything now, both on the farm and off.

I still call Alfred when I'm in over my head. I still hire help for the big, complicated projects like installing electrical lines, major structural work, anything that requires actual expertise. But I no longer feel useless. And I no longer pay people to build my IKEA furniture.

In fact, these days, I make a lot of our furniture myself.

The very desk I used to write this book? I built it with my own hands. It's not perfect. The edges aren't quite square, and if you look too closely, you'll see every mistake I made along the way.

But it's sturdy and it does its job.

And unlike that IKEA chair, it never once made me question my entire sense of self.

CHAPTER 12

WHY COUNTRY FOLK HATE CITY PEOPLE

I remember walking into the forest, wondering if I was crazy. Here I was with four strangers, three men and a boy. Chains clanked softly against the metal of their firearms. We moved deeper and deeper into the forest. At this point, I realized that nobody outside this group knew my whereabouts. I started to feel like I was going to puke, but I kept making conversation with the older man as a way to stabilize my nerves and draw out more information.

"How far away would you say they are right now?" I asked.

A handheld GPS system beeped as the old man, who looked somewhat like a hard-luck Santa Claus, glanced at it.

"Two or three hundred yards maybe," declared one of the other men.

The eerie chorus of hounds howling in the distance broke through the forest's usual natural symphony.

"Look, people have been using hounds since before the Romans," he said as he cut through the underbrush of the cedar swamp.

"I appreciate that. People had slaves back when there were Romans. Doesn't mean you should be having slavery, right?" I shot back.

"Sometimes, I wish we did," he grumbled wistfully.

At that moment, a cold realization washed over me: What the hell am I doing out here? Who exactly were these people? How did I end up in such a precarious situation?

About half an hour earlier, the old man with the white beard had pulled up in my driveway. He jumped out of his truck and immediately produced a scrap of paper cut in the shape of a business card with stock clip art of a bear weakly printed on it.

"This is who I am and what I'm about."

A bear hound hunter. While more common in the South than New England, hound hunters release their packs of dogs to sniff out prey across the countryside, going where the bear, coyote, raccoon, or whatever game they are chasing may go. Oftentimes, the hound hunters will be miles away from their dogs, following the little electronic blip on an LCD screen produced by the dog's GPS collar. The man's dogs had chased a bear right into the dense cedar swamp of my 160-acre farm. It wasn't the first time. My farm seemed to attract hound pursuits magnetically.

Just a few weeks earlier, my wife and I had been woken up in the middle of the night by strange men carrying guns, chasing their hounds through our dooryard. Toby Dog had gone nuts. My ducks and geese were honking and quacking in terror. Shouting, threats, and harsh words were exchanged. My wife called the state troopers. The strange people with the dogs had already left when they showed up. But to my shock, the state troopers and game warden informed me what the people with the dogs had done was perfectly legal.

Now here I was again. Different hunters, different dogs, but the same frustrating scenario playing out in daylight. Standing in the forest, I took a deep breath, refocusing on the immediate concern: making sure the hunters retrieved their hounds but let the bear go free.

"You can get your dogs," I told them firmly, "but the bear goes free."

He hesitated, then nodded reluctantly. I lifted my GoPro and followed, capturing every tense moment.

I later posted the video of the encounter on the internet, and it went viral, bringing a lot of attention and discussion on the topic of hound hunting to the state of Vermont and across New England.

This incident marked the first major time that I, a city person, had run afoul of the country folk I now lived with. Three years earlier, my wife and I had left DC and moved to an old farm in the second least populated state in the United States of America, in the poorest and least populated corner of that small state, in a town of roughly seven hundred people. The town used to be a dairy town, but that industry was slowly dying out. And with it, a way of life was also dying out.

I've understood that the divide between rural and urban life is deep and historical, stretching back to ancient Europe. In those days, a peasant could be put to death for chopping down a tree in the king's forest. The resentment toward city folks from country dwellers is deeply rooted. As I reflect on these tensions, I can't help but think that if America ever faces another civil war, this rural-urban divide will be at its heart.

People from urban areas often view the countryside with a mixture of romanticism and disdain. To them, rural landscapes are either the pastoral scenes of idyllic novels or bleak stretches of cultural desert populated by folks who lack sophistication. Perhaps inherited from books, movies, or TV, there's a notion that rural folks live "simpler" lives, but the simplicity is often mistaken for naivety.

It's not that rural life is less complex. It's just different. Urbanites might chuckle at the "country bumpkin" who is bewildered by city traffic and skyscrapers, seeing it as a sign of a broader disconnect with the modern world. But those same urban dwellers are often terrified by the streetlight-less darkness that comes to the countryside each night.

Conversely, many in rural areas view city dwellers through a lens tinted with skepticism and a bit of resentment. They see the urbanite as someone who is perpetually in a hurry, someone who lives life at a superficial level, always connected to the latest technology but deeply disconnected from the natural world and traditional values. There's a perception of entitlement too, as if the city folk expect the world to cater to their every whim, whether in service or convenience. This stereotype is bolstered by stories of city residents who move to rural areas and then complain about the early morning crow of a rooster or the smell of manure.

These stereotypes don't come out of nowhere. They're stitched into history, woven from the stark, undeniable differences between rural and urban life. Cities? They've always been the beating hearts of commerce, technology, and cultural exchange, restless, ambitious, forever shifting under the weight of the next big thing. They move fast because they have to. The economy demands it. The people demand it. The whole machine runs on progress, or at least the illusion of it.

Rural places, though? Different story. Out here, change doesn't come in bold leaps but in slow, deliberate turns. Seasons passing, land being worked, the same roads leading to the same places. Agriculture, resource industries, the kind of work that doesn't stop just because the world decided to move faster. There's a rhythm to it, a weight, a sense that tradition isn't just nostalgia. It's survival.

Put those two worlds side by side, and of course, they create drastically different ways of life. Not better or worse. Just different speeds, different priorities, different definitions of what it means to move forward.

This cultural chasm isn't unique to modern America. It's just history remixed. The same old tension, dressed up in contemporary clothes. Long before highways and skyscrapers, before Wi-Fi and Whole Foods, there was the old-European split: the old money landed gentry versus the new money urban merchant class.

In feudal times, power wasn't measured solely by money, but by control over land and the people who worked it. Nobles managed large estates, their influence deeply tied to agricultural cycles and entrenched privileges. Meanwhile, in towns and cities, merchants, traders, and artisans built their wealth differently, accumulating prosperity through commerce and craft rather than direct control of the land. Their growing economic influence began to challenge traditional power structures, causing friction between established rural aristocracies and rising urban classes.

In my opinion, that's where a lasting tension took root. Rural elites saw city dwellers as disruptive, ambitious outsiders threatening the established order, while urban communities viewed rural lords as entrenched and resistant to progress.

Sound familiar? It should. Because while times have changed, echoes of this historical divide persist today, transformed into our modern urban-rural tensions.

The rural/urban conflict has also been at the very heart of the conflict in the United States since its founding. Back in the 1790s, the earliest days of USA, Inc., the Whiskey Rebellion sprang from the fertile fields of frustrated farmers. Looking for ways to raise revenue, the federal government placed an excise tax on whiskey, introduced as part of Alexander Hamilton's strategy to manage the national debt and solidify federal economic clout. For the farmers of western Pennsylvania, whiskey wasn't just a commodity but a vital necessity. Corn, their primary crop, was difficult to transport across the rugged Appalachians. By distilling it into whiskey, they reduced its bulk and increased its worth, making the tax feel like a threat to their very survival. To these rural distillers, the tax seemed an imposition by distant urban elites who seemed not to feel its sting, widening the gap between the rural producers and the urban financiers.

As the century wore on, the balance of power in America lurched toward the cities. Industrialization, immigration, and the sheer gravitational pull of economic opportunity turned urban centers into the hubs of influence, shifting wealth, culture, and politics away from the fields and into the crowded, chaotic streets. If you were the kind of person who thought about progress in terms of factories and banks rather than fields of wheat, this was good news.

Alexander Hamilton certainly thought so. He took one look at the churning energy of urban commerce and saw the future: strong federal governance, a diversified economy, a nation that made things, moved money, and didn't waste its time sweating in the sun. Thomas Jefferson, on the other hand, clung to the pastoral ideal: a nation of noble farmers, untainted by the corruption of city life, whose independence and self-reliance would keep democracy pure. If Hamilton saw cities as progress, Jefferson saw them as contagion, a place where ambition curdled into greed and where the simple virtues of agrarian life went to die.

When the framers designed the US government, they attempted a fragile balancing act by constructing a bicameral legislature to manage the tension between densely populated states and rural, less populous ones. The Senate provided equal power to each state, while the House of Representatives reflected population size. It was a pragmatic but deeply flawed compromise, shaped in part by efforts to protect the institution of chattel slavery.

Today, these flaws remain starkly apparent: Mechanisms like the Electoral College and equal Senate representation can leave voters in heavily populated states like California or Texas with diminished political influence compared to citizens in sparsely populated places like Vermont or Wyoming. It's not a question of one side being inherently better or worse, but rather the legacy of an imperfect, morally compromised compromise embedded deeply in our democratic structure.

As cities expanded, rural America dug in. The rise of urban wealth felt like a hostile takeover, a centralization of power that made it clear who the economy was working for. When the federal government rolled out initiatives like the national bank, it only confirmed their suspicions: Urban elites were making the rules, and they weren't making them for farmers. The tension wasn't just economic; it was cultural, ideological, a slow-burning feud that never really resolved.

Historically, cities have depended on rural areas for vast resources. Food, raw materials, and labor flow from the countryside to urban centers, fueling the industries and lifestyles that define city living. From the coal mines of Appalachia to the vast cornfields of the Midwest, rural America has been the cornerstone upon which urban prosperity has been built. Yet, this dependency often hasn't translated into equitable benefits for rural communities.

Urban centers, teeming with people and their taxable incomes, send money back out to rural areas, funding roads, hospitals, schools—basic infrastructure that many of those communities couldn't swing on their own. Even with that back-and-forth, rural areas often get left behind, facing fewer opportunities, less investment, and too little of the economic diversity that cities take for granted. Like most things in

America, it's complicated, and nobody is ever totally happy with the deal they're getting.

The urban-rural divide isn't just about economics or politics. It's also a cultural problem. The people who tell America's stories live in cities, work in cities, and, for the most part, think like city people. And when they turn their gaze toward rural life, what they see tends to fall into two neat categories: quaint and folksy, or backwards and doomed.

TV and movies love a nostalgic, small-town setting, as long as it's frozen in time, populated by either kindly old shopkeepers or rugged individualists who speak exclusively in plainspoken wisdom. The moment rural America insists on existing in the present, things get more complicated. More often, the dominant narrative frames it as out of sync, out of step, or just plain out of ideas.

And that portrayal matters. When rural communities turn on the TV and see themselves reduced to either simple country folk or cautionary tales, the message is clear: They're not the ones writing the script. The people who shape the national conversation aren't from there, don't live there, and don't particularly care whether they're getting it right.

The educational divide further entrenches this disparity. Higher education institutions and opportunities are predominantly located in urban areas, drawing young, educated individuals away from rural communities, a phenomenon known as "brain drain." This migration depletes the rural workforce of its potential leaders and innovators. It ensures that the benefits of education, such as increased earning potential and enhanced social mobility, are enjoyed predominantly by urban populations.

The result? Resentment. The kind that doesn't just fade after one bad election cycle. Rural voters don't just see policy when they look at urban-led initiatives. They see a threat, a reminder that the people making decisions don't have to live with the consequences.

It's taken me years to appreciate all this complexity, and even longer to see just how deep this divide runs. But the more clearly I see it, the more convinced I become that it's one of the defining fault lines in America today.

And what really bothers me is how few people in positions of power are working to bridge that gap. Instead, too many politicians see it as an opportunity, an easy lever to pull when they need votes or a wedge to drive deeper divisions when it serves their purpose. Rather than trying to help communities understand each other, they lean into stereotypes and grievances, widening the chasm between rural and urban America, and leaving us all a little worse off.

One of the odd quirks of being a farmer who puts their farm on the internet is that I regularly get hate mail. Sometimes, it's as benign as rude comments on a video post. Other times, it comes in the form of pages of longhand-written letters that denounce me and everything I stand for. But when the hate mail comes in (and it definitely comes in!) there's often a common and consistent refrain: Look at this city guy trying to change country life.

"City folk just can't help but carry the city with them," they say.

"Stop trying to Californiacize the country," say others.

And finally, from one particularly creative emailer: *"Look at this cuck Jew playing Green Acres. Only thing your growing is Trust Fund."*

It's a bitter pill. I've stood by my views unapologetically, but I'm beginning to see there's more to the backlash than just a resistance to change. There's a sense, maybe, that in fighting for what I believe is right, I'm also trying to "citify" the countryside. That was never my intent, but it's worth asking myself if maybe that is true.

While I moved to a farm in the middle of nowhere to become a farmer, that doesn't mean that that choice made me surrender my values, likes, and dislikes. And while I might be willing to embrace the smell of spring manure spread on the neighbor's pasture, I might also have a strong distaste for strangers chasing their dogs wearing GPS collars for miles while they attempt to hunt bears on my posted land. I think figuring out where to draw the line can be complex. Conflicts between people born in a place versus transplants will be inevitable.

But here's another important question: Just because you're new, does that mean your ideas are less important? Absolutely not! Everyone's thoughts and opinions should matter. It's important for both old-timers and newcomers to listen to each other. That way, everyone can help make the town a better place to live. In my opinion, a healthy community isn't static and frozen in time. It's constantly evolving and changing. It's a living organism.

This transplant-versus-native tension has only sharpened in recent years, accelerated by the global shifts brought on by the COVID-19 pandemic. The ability to work remotely has drawn more city dwellers into rural settings, armed with city-fueled bank accounts and a taste for larger plots of land, altering the demographic and economic landscape of places like Vermont.

From the perspective of a city dweller, the opportunity is almost absurd. Back when we were house hunting in New York, we couldn't even land a one-bedroom apartment despite making offers well above asking. The monthly payments alone would've rivaled the average household income in most small towns. By contrast, when we found our farm—160 acres, massive outbuildings, and a seven-bedroom house—we paid roughly 1/3 of the purchase price of our tiny DC rowhouse. That's not a small difference. That's an entirely different financial reality. For someone used to urban prices, it's like buying a football field for the cost of a parking space.

But when you look at the transaction from the perspective of a rural person, that's downright terrifying.

Imagine you've lived in the same cozy little town all your life. Your family has been there for generations. You know everyone's name and stories. The fields, the woods, and the town square hold countless childhood memories, and every face you see is familiar. It's a tight-knit community where neighbors look out for each other, and every new face is noticed.

Now, picture this: Suddenly, new folks are moving in. Not just one family, but many, coming from the big cities. They start buying up land, much more than they need, and for prices that seem like fortunes to you. Houses that once sold for what you thought was a fair price are now selling for double or triple that. It feels like your quiet hometown is becoming something you hardly recognize overnight.

These newcomers can be nice enough, but they don't know the rhythms of life here. They don't greet you the same way your native neighbors did, and they talk about changes they want to make, changes that don't make sense to you. They see empty fields and think of building new homes or businesses, not realizing those fields were where you learned to ride your bike or where the town's annual picnic has been held for decades.

It's like watching your home slowly erased and redrawn into something unfamiliar. You feel fear and sadness as you wonder if the town's old ways will be remembered. You worry whether the new residents will ever understand the value of preserving the heritage that makes this place special or if they'll change everything to look more like the busy, crowded cities they left behind.

And yet, people move. They always have. Migration is part of what shapes communities, for better or worse. It's not inherently wrong for someone to want a quieter life, more space, or a stronger connection to the land. But the key is in how you arrive. You can move to a place without trying to remake it. You can appreciate the peace and beauty of rural life without needing it to come with artisan lattes and round-the-clock delivery. The challenge and the responsibility is learning how to live in a place without trying to own it.

The viral video of the bear hound hunters marked a turning point in an increasingly tense situation. Despite promises from both sides to open up a dialogue, genuine communication never materialized. My later videos featuring other conflicts with bear hound hunters only seemed to intensify the conflict. When I posted another video in the fall of 2022, the reaction was swift and hostile. Someone featured in the video shared it on Facebook, prompting commenters to discuss "paying me a visit," ensuring my security cameras were disabled, and ominously posting fire emojis.

A week later, in the middle of the night, some people came to the farm. They disconnected the road-facing security camera and shot road flares at the window of my hay-filled barn. The shots missed the open barn window by inches. If they had been on target, the barn would have almost certainly burned down. The police never caught the folks who did it, and the tensions with the hound hunters remain high to this day.

CHAPTER 13

A FARMER WHO TELLS STORIES

Most mornings on the farm, I'll wake up early and head outside first thing. I need to start the day's chores. That's standard for most farmers. But unlike most farmers, I don't just bring buckets and hoses. I bring an entire camera crew. Well, a one-person, self-operated crew made up of a constantly changing arsenal of cameras: expensive DSLRs with fancy lenses, cheap and indestructible action cameras, AI-powered drones, and even a tiny little camera that I can mount on my ducks and cats.

To me, the cameras are as much a part of farm work as my trusty Leatherman. I know this is not normal.

The fantasy of farming is intoxicating. You picture the sunrise over misty fields, the rhythmic hum of crickets, and the quiet satisfaction of hands in the soil. And for a while, I bought into that fantasy completely. A kind of agrarian utopia, where my corporate past faded behind me, replaced by honest, tangible work with my hands and the land.

I wanted this life. But from the start, there was an unease I couldn't shake. And all it takes is a quick scroll through the comments on my latest video where I can have that sense of unease confirmed. Nine out of ten comments on most of my videos are positive, supportive, and wonderful. But it's the one out of ten you remember that may send you into a tailspin.

I tell myself I don't care. These are just trolls shouting from the dark abyss of the internet. I'm glad my life isn't so horrible that I would spend hours of it watching videos on the internet that anger me. But the truth is, I do care. Because deep down, I'm not so sure they're wrong.

On paper, I am a farmer. I have the land. I have the animals. I do the work. I make money from that work. But there's this nagging, gnawing insecurity I can't shake. Unlike generations of farmers before me, I'm not keeping the farm afloat by selling what I produce. But I also don't have to keep my farm alive by working an off-farm job like 92 percent of farmers in America do. I'm keeping it alive by documenting what I'm doing.

What kind of farmer makes more money off *talking* about farming than *actually* farming?

At first, the videos were meant to be just marketing tools to sell duck eggs. If I wanted to be New England's most prominent natural-duck-egg producer, I would need a social media following to market to. But soon, the social media income blew past farm income by a wide margin. And that bothers me. It makes me feel like an impostor because I'm not living purely off farming. I'm living off the attention economy.

Every time I pick up the camera instead of whacking weeds or fixing fences, I wonder if farming has truly become secondary. I tell myself farming is still the priority, that the videos are just a side gig. But if that's true, why do I feel more pressure to hit my upload schedule than to increase my cattle herd size? Why do I spend more time editing videos than staffing market booths?

The answers to those questions make me uncomfortable.

For most of human history, farmers survived off their ability to produce. You raise crops, tend livestock, sell what you can, and hope the margins keep you afloat. The idea that someone can "farm" while making most of their income through video views, ad revenue, and sponsorships feels absurd even to me, even as I'm living it.

So, I bury the conflict.

I tell myself that I'm just modernizing the farm business. That old-school farmers had Farm Aid concerts and government subsidies, and I have YouTube. That I'm still doing the work, just finding a new way to make it pay.

And yet, the voice in the back of my head keeps whispering, *You don't belong here.* This isn't just about trolls on the internet. This is about me wrestling with what it means to be a *real* farmer. And the longer I spend behind the camera, the harder that question is to answer. But to really understand this tension, you need to understand how the farm actually makes money because, like it or not, storytelling is the backbone of this operation.

Every year, I make a video breaking down my farm's finances. Line by line, dollar by dollar, I lay it all out: what I spent, what I earned, what worked, and what didn't. But I keep the focus of these videos on just my traditional farming operations: eggs, meat, trees, and so on. And every single year, the response is the same. Some people appreciate the transparency and detail. Some people are shocked by the numbers. And a whole lot of people get mad.

Because when they see the breakdown, it becomes obvious that my farm doesn't make most of its money from farming.

It makes money from content creation.

Here's how it works:

1. **Ad Revenue**: Every time someone watches one of my YouTube or Facebook videos, an ad plays. The social media platforms pay me a portion of that ad revenue. This is the most passive form of income I have. Once a video is out there, it keeps generating money as long as people keep watching it. Some videos might make a few bucks a month. Others rack up millions of views and turn into significant revenue streams.
2. **Sponsorships**: Brands pay me to feature their products in my videos. But I'm picky. For every hundred offers I get, I might accept one. It has to be something I genuinely use, believe

in, and feel comfortable recommending. Otherwise, it's not worth it.

3. **Merchandise Sales**: People who watch my videos buy Gold Shaw Farm hats, shirts, and stickers. Merch isn't a huge part of my revenue, but it's a steady stream. And more importantly, it turns viewers into a real community. When I see someone at a farmers' market, a grocery store, or an airport wearing a Gold Shaw Farm hat, it reminds me that this is bigger than just me.
4. **Book Sales**: Writing has always been a part of my life, and now it's a part of my farm business. Books, like this one, offer another revenue stream that isn't dependent on fickle social media platform algorithms or the whims of advertisers.
5. **Farm Product Sales**: Yes, I do sell farm products. Right now, that includes mostly meat and trees. But if you were to rank my income streams by percentage, farm product sales are at the bottom of the list.

When you add it all up, most of my income comes from storytelling, not from farming in the traditional sense. This is where the tension really starts. On the one hand, I'm not alone. A huge number of farmers, especially small-scale farmers, rely on off-farm income. The USDA reports that roughly 90 percent of American farm households earn the bulk of their income from somewhere other than the family farm. Compared to that, making a living through storytelling doesn't seem that unusual.

But on the other hand, it feels different.

If YouTube, TikTok, Facebook, and Instagram all disappeared tomorrow, my farm wouldn't financially survive in its current form. That's the reality. And no matter how much I try to justify it, that fact still nags at me.

Traditional farmers measure success in bushels, pounds, gallons. I measure it in views, engagement, watch time. Some days, I worry that I'm just chasing numbers on a screen, that I've built something fragile, dependent on algorithms and fleeting internet attention.

And yet, this is real. The farm is real. The work I do is real. I'm not fabricating anything. I'm not staging moments for the camera or pretending to be something I'm not. I'm documenting my life, my farm, my struggles, and my failures.

And people watch. A lot of people.

Which is why, despite all my doubts, I keep picking up the camera. That and the money. You can't forget the money.

When I was a kid, the idea of making a living by sharing your life online wasn't just improbable. It was unimaginable. The internet was barely functional, and the concept of social media didn't exist. Even when I graduated college, no one considered "content creator" a real job.

Yet here I am, working a job that didn't exist a generation ago. And, remarkably, it's a job that most kids today *want.*

Research shows that a massive percentage of Gen Alpha (kids born between 2010 and 2025) dream of becoming YouTubers, live streamers, or influencers. A 2019 survey found that more children aspired to be YouTubers than astronauts. Newer data suggests that nearly 30 percent of Gen Alpha wants to be content creators when they grow up. They see people like me documenting their lives, telling their stories, and building businesses from it, and they think, *that's the dream.*

It's surreal to realize that my career, something that still feels experimental and precarious to me, is now considered aspirational. Because when I first started making videos, I didn't quite understand that videos about farm animals could generate real income. I wasn't thinking about sponsorships, ad revenue, or merchandise.

I was just trying to sell duck eggs.

When I moved to Vermont to start my farm, I needed a marketing strategy. I wasn't a fifth-generation farmer with built-in customers and a long-standing reputation. I was a guy with some land, some ducks and a dream. So, I started making videos to document what I was doing, hoping that maybe, just maybe, someone would be interested enough to buy my products.

But something unexpected happened.

People weren't just interested in *buying* from me. They were interested in *watching* me.

The first time I got a check from YouTube, it felt like a fluke. A couple hundred bucks for doing something I enjoyed? Cool. But as more people found my videos, those checks got bigger. Soon, I wasn't just making *some* money from content creation. I was making *most* of my money from it.

It was a strange realization. Farming was supposed to be my livelihood. But suddenly, my financial stability didn't depend on the number of ducks I had. It relied on the number of people watching my videos.

Without even realizing it, I had stepped into the world of influencer marketing and digital storytelling. And that world had rules of its own.

I was lucky. If I had tried this even ten years earlier or seven years later, I don't think it would have worked.

There was a time when creating high-quality video content required thousands of dollars in equipment, specialized skills, and industry connections. You needed expensive cameras, professional editing software, and a level of technical expertise that made filmmaking exclusive. Heck, I learned how to edit using razor blades and Scotch tape. But now? I could create a full-length feature film using only my cell phone.

The shift from heavy, expensive camera gear to affordable DSLRs, GoPros, and even AI-powered drones has made high-quality video production accessible to just about anyone. And the rise of social media platforms like YouTube, TikTok, and Instagram has transformed how stories are told and consumed.

Attention spans are shorter, competition is fiercer, and AI slop is everywhere. While it is true that I earn most of my living in a way that didn't exist a generation ago, I have strong doubts that the current model will still be viable a decade from now.

I didn't set out to become a digital storyteller. I just picked up a camera and started filming my life. But whether I realized it or not, I had entered a space evolving at a breakneck speed, where content wasn't just something people watched. It was something they *valued.* It was a type of work that I needed to do.

But even before YouTube or Instagram, the tension between storytelling and authenticity existed. Just look at Helen and Scott Nearing. Their book, *The Good Life*, had become something of a Bible to the back-to-the-land movement of the 1960s and 1970s. The Nearings were intellectuals who, in the 1930s, left behind city life to homestead in Vermont. They built their own stone houses, grew their own food, and constructed an image of radical self-sufficiency that inspired generations of homesteaders, environmentalists, and idealists—including me.

The problem? The Nearings could only live their version of "The Good Life" because they had money. Scott Nearing had inherited large sums of wealth. The couple supplemented their farming income with book sales, speaking engagements, and, later, a revolving door of eager disciples who paid for the privilege of working on their farm. They weren't truly living off the land. They were living off a story about living off the land.

When I first read their work, I didn't see that contradiction. I saw a blueprint for the life I wanted. But once I started farming, the reality of running a modern small farm quickly set in. It didn't take long to realize that selling duck eggs alone wouldn't keep the lights on. I still worked my day job, so that's how I was really paying the bills. But if I was ever going to make farming my day job, I would need to win the lottery.

That's when I started looking around and realizing something: The Nearings weren't the only ones who had built a life from the story of self-sufficiency rather than the practice of it.

Scroll through Instagram, YouTube, or TikTok, and you'll find countless modern homesteaders showcasing their idyllic farm lives. Perfectly framed shots of morning chores, glistening produce fresh from the earth, and rustic farmhouses that seem plucked from a Norman Rockwell painting. What you don't always see are the YouTube ad revenue checks, the sponsorship deals, the off-camera reliance on store-bought feed, or the fact that many of these so-called self-sufficient farmers are making most of their income from content creation, not farming.

Social media has turned the dream of self-sufficiency into a curated spectacle. You scroll through your feed and see a sun-drenched homestead where someone in a perfectly worn flannel gathers eggs from a picturesque wooden coop. A sourdough loaf cools on a rustic farmhouse table. A bearded guy in overalls plants rows of heirloom tomatoes in soil that looks like something out of a glossy lifestyle magazine. It's serene, it's wholesome, and it's deeply aspirational. Heck, it might even have been these idealized versions of farm life that motivated you to buy this very book.

And for the average person who is stuck in a job they hate, living in a city they feel disconnected from, it's intoxicating. They look at those images and think, *That's what I want. That's what I need. If I could just get some land, raise some animals, and grow my own food, life would be simple. Life would be good.*

But what they don't see—what they *can't* see—is everything outside the frame. They don't see the long hours editing videos to make those shots look cinematic, or the brand sponsorships that are funding the farm. Those beautiful, thriving homesteads often make more money from selling the *idea* of farming than from *actual* farming. (Won't somebody think about the influencers!)

Farm life, in general, is messy, exhausting, and often brutal. It's trudging through ankle-deep mud in freezing rain to fix a fence that a panicked cow plowed through. It's the sick feeling in your gut when you find an ill calf cold and lifeless, despite doing everything you could. It's butchering day, when the animals you've hatched yourself are taking their final walk, and you have to reconcile the ethics of meat with the blood on your hands. But when you sand down the sharp edges of farm life and focus only on the Instagram-friendly aesthetic moments, you're leading people down a dangerous, often misunderstood path.

That's why transparency matters.

If people are going to romanticize farming—and they always will—then at the very least, they deserve the full story. They deserve to know that running a small farm isn't just about fresh air and homegrown meals. It's about figuring out how to make ends meet in an industry where margins are razor thin. It's about using every tool, including social media, to create a sustainable life.

And if you're honest about that? If you pull back the curtain and say, *Hey, this is how it actually works*? Then maybe, just maybe, you're helping people instead of misleading them.

It's not that there's anything inherently wrong with that. These days, I'm doing the same thing. The difference is whether or not you're honest about it.

For a long time, I wrestled with the idea that making money from storytelling somehow made me less of a farmer. The comments online—"you're just a YouTuber, not a real farmer"—burrowed into my brain because, on some level, I feared they were right. But the more I thought about it, the more I realized that what makes something dishonest isn't the act of storytelling. It's *how* you tell the story.

Maybe that's exactly what I was meant to do.

The first real story I ever told wasn't a farm story. It wasn't even a video. It was a comic book.

I must have been around seven or eight when I made my first one. I remember sitting on the floor with a stack of printer paper, sketching out panels with a ballpoint pen, creating a world far bigger than the one I lived in. By the time I was in high school, I was making and distributing my own little mini-comics and zines. It became the thing I was known for, the thing I loved doing the most.

I was never a great artist. Even now, I'd say my drawing skills are mediocre at best. But storytelling? That was different. I could take an idea and bring it to life, laying it out in a way that kept people interested, made them laugh, or made them think. I had a knack for structuring a narrative, for figuring out how to make a story *work*.

And that's what I'm still doing today.

When I got older, my storytelling took on different forms. I studied animation in college. My first real job out of school was at a small corporate video company in Connecticut, shooting and editing training and promotional videos for insurance firms and manufacturing companies. It wasn't glamorous, but it taught me the technical side of things. How to shoot, how to edit, how to make something *watchable.*

But something about it felt…off.

I'd spend hours perfecting shots of conference rooms and safety demonstrations, but the stories weren't *mine*. They weren't stories I cared about. And that itch, the one I'd felt as a kid making comics, never really went away.

That itch is what led me here.

When I started my farm, I thought I was leaving my past life behind. No more office work. No more video editing. Just farming. Just real, tangible work.

But almost immediately, I felt the pull to document it. Not just because I needed a way to market my duck eggs, but because I couldn't *not* tell the story. I found myself structuring my farm life the way I used to structure comics and videos. Beginning, middle, end. Setup, tension, resolution. The struggle to start the farm became a narrative.

And the more I leaned into that, the more I realized: I wasn't just a farmer. I was a storyteller who farmed.

For a long time, I wrestled with that identity. Was I a *real* farmer if I made more money telling stories about farming than actually farming?

But when I look back, seeing that kid on the floor drawing comics, that teenager running off Xerox copies of homemade zines, that corporate video editor cutting together client testimonials, I find the pattern.

I was *always* a storyteller. Farming didn't change that. It just gave me a better story to tell. And for the first time, I started to wonder: Maybe this wasn't an accident. Maybe this is exactly what I was meant to do.

As I climbed the ranks in the corporate world, I found myself moving away from cameras and editing software and deeper into marketing and communications. One big break came when I landed a job writing speeches and presentations for a Fortune 100 insurance company's CEO. I hadn't planned on becoming a speechwriter, but when the company's previous writer left, I volunteered to step in. At first, I was just faking my way through it, but over time, I figured out the rhythm of structuring a story so it would land with maximum impact on an audience.

I kept moving deeper into marketing, working in different corporate roles, taking on bigger projects. And with each promotion, I became more financially successful, but also more creatively unfulfilled. The further I got from hands-on storytelling, the more I felt like I was just pushing corporate jargon around. I was good at it, but I didn't love it.

By the time I was in my mid-thirties, living in Washington, DC, I had a comfortable life but an ever-growing sense that I was wasting my time. I still had creative projects on the back burner, including a script for a film about Homo sapiens fighting Neanderthals and a graphic novel tracking the history of a single farm from 1776 to the present day. But none of them ever went anywhere. I'd pick them up on a Sunday afternoon, tinker with them for a few hours, then let them gather dust again.

That's when I discovered farming videos on YouTube. Initially, I was just curious about farming as a hobby, but as I watched people share their small-farm lives, I realized I already possessed many of the skills they used, skills I hadn't tapped into for years. I knew storytelling, shooting, editing, and how to captivate an audience.

Looking back, it makes perfect sense. I've always been a storyteller. The mediums evolved from comics, to animation, to corporate marketing, and eventually YouTube, but the core impulse never changed.

The only difference now?

For the first time, I was telling stories of a subject matter I deeply cared about.

On the surface, it might seem like I just turn on the camera and capture whatever is happening. Just a guy documenting his life, nothing more. But at some point, I had to admit the truth: There's a difference between just filming what happens and actually *crafting* a story. And whether I wanted to or not, I had been crafting stories my whole life.

Telling stories on social media isn't the same as writing a novel or making a documentary. It's faster, more disposable, and shaped by algorithms that reward engagement over depth. But at their core, the rules of storytelling don't change. People still want to feel something. They still need a reason to care.

Nobody is required to watch your content. Nobody owes you their attention. This was a hard lesson for me at first. I'd spend hours working on a video, thinking, *This is good. This is interesting.* Then I'd post it, and…nothing. People clicked away. In the early days it was only a couple of friends and Allison's dad who watched them. The algorithm buried it.

It wasn't until I really started *studying* what worked that I saw the pattern. The videos people responded to weren't necessarily the most beautiful, the most polished, or even the most well-shot. They were the ones that made them *feel* something. That's the trick with storytelling. People don't just want information. They want to be engaged. They want to care. They want to see a challenge unfold and feel like they're right there with you, experiencing it in real time.

For a long time, my videos were just that: videos. I'd film the ducks, talk about the trees, show my daily work, and they were fine. But they weren't *compelling*. The turning point came when I started thinking about my videos the same way I used to think about comic books and marketing campaigns. Every good story has a structure: a setup, a conflict, and a resolution. Without conflict, there's no reason to keep watching.

A video about daily chores? Boring.

A video about trying to fix a frozen water line in the dead of winter while my ducks stare at me like I'm an idiot? That's a *story*.

If you want people to stay engaged, you have to make them wonder, *What happens next?* That's why I always look for the tension point. The question that keeps people watching. The problem that needs to be solved.

Here's the paradox of social media storytelling: The best content feels effortless, but it never is.

When I first started making videos, I thought people just wanted raw, unfiltered reality. No planning, no scripting, just "real life." And while *some* people connect with that, the reality is that most audiences don't stick around unless they're *entertained.*

That doesn't mean being fake. It doesn't mean staging things. It just means *shaping* reality into a story. A good documentary filmmaker doesn't make things up. But they *choose* what moments to highlight. They frame things in a way that makes you feel something.

I've had to learn how to balance this. If I don't make the effort to shape my videos into compelling narratives, people tune out. But if I lean too far into "producing" my life, it stops feeling real to me and to my audience.

The trick is to find the balance. To document *honestly*, but structure *intentionally*.

One of the strangest things about being a storytelling farmer is that my life is both completely public and completely private at the same time. People who watch my videos feel like they know me. They see my farm, my animals, my struggles, and my wins. But they don't see *everything*.

And that's intentional.

Early on, I struggled with this. I felt like I *owed* my audience full transparency. If I wasn't showing every single part of my farm, every financial decision, every business move, every personal moment, was I being dishonest? Was I hiding something?

But over time, I realized that's an impossible standard. No one sees *everything*. No one should. Storytelling isn't about dumping your entire life online. It's about crafting something meaningful out of the parts that matter most. Now, I have clear boundaries. There are things I *don't* share. Not because I'm being deceptive but because not everything needs to be content. Some things are just...life.

I frequently get messages from people who want to start making their own videos but don't know how. They're worried about the gear.

The editing. The algorithm. They want it to be *perfect.* It won't be. Your first videos will suck. And that's OK.

My first videos were awkward. My first edits were bad. My first attempts at storytelling were clumsy. The only way I got better was by doing it. Over and over. If you wait until you feel "ready," you'll never start.

The best way to learn storytelling is to *tell stories.*

Every story, no matter how big or small, follows a pattern. You start in one place. You want something. You leave behind what's familiar to chase it. You struggle. You fail. You learn. And if you're lucky, you come out the other side changed.

That's not just a storytelling framework. That's life.

I started this journey believing I was just a guy who wanted to farm. But the truth is, I was never *just* a farmer. I was never *just* a corporate marketing guy. I was never *just* a kid making comics on his bedroom floor. I was all those things, all at once, slowly adding up to the person I am today.

A storyteller. A farmer. Both.

For years, I struggled with the idea that if I wasn't making my living *directly* from the land, I wasn't a "real" farmer. But the longer I've done this, the more I realize that storytelling isn't separate from farming. It's a part of it.

A farmer who can't sell their story struggles to sell their products. A farmer who can't explain their work has a harder time building a community. A farmer who shares their life honestly and fully helps others see that they, too, can carve out a different kind of existence. So now, I don't fight it anymore. I embrace it.

I may not measure my success in bushels, gallons, or acres. But I still grow something. I still raise something. And if I do my job right, I leave something behind that matters: A story worth telling.

CHAPTER 14

NOT THE DOG I WANTED (BUT THE DOG I NEEDED)

The laptop screen glared at me. The venomous words cast a cold, stark light illuminating the room. Abby Dog's breeder, Bertha, had just published another scathing blog post about me. This time, it felt personal. An open condemnation of me as a farmer, a caretaker, and a person.

"It's tragic to see an inexperienced farmer ruin a dog like Abby. A true livestock guardian dog deserves more than to be wasted on someone playing pretend at farm life."

The sentence felt like a blade, sharpened specifically for my throat. My face burned with humiliation as I reread the accusations: that I was incompetent, that Abby Dog's every failure as a guardian was my fault. The breeder didn't stop there. She insinuated that my entire farm was a joke, that I had no business raising animals, let alone training a dog. She questioned my commitment, my ability, and even my intentions.

"Perhaps if the farmer spent less time making videos and more time actually learning to care for a farm, Abby Dog wouldn't be struggling."

I slammed the laptop shut, the words ricocheting in my head like shrapnel.

Outside, the farm was chaos incarnate. Chickens were sprinting like extras in *Mothra vs. Godzilla*, while Abby tore through them like a velociraptor with boundary issues. Feathers floated down like sarcastic confetti. Ever the dignified professional, Toby Dog barked at Abby with the exasperation of a babysitter dealing with a sugar-high toddler. His side-eye was so sharp it could cut wire fencing.

"Abby Dog! *No*!" I shouted, my voice cracking like I was asking out a junior prom date.

Abby Dog skidded to a halt, her thick paws throwing up a spray of dirt. She turned to face me, her wide brown eyes filled with glee and a hint of defiance.

Abby Dog was meant to complement Toby Dog: to be his partner in protecting the farm, to help raise the defenses against predators, and to share the workload. Heck, I even had initially harbored dreams of breeding the two of them and creating an entire line of Maremma Sheep Dogs coming from our farm. Instead, Abby Dog was a whirlwind of unchecked energy, disrupting everything Toby Dog and I had built.

And it wasn't just the breeder who had noticed. Online critics had joined the chorus, commenting on every mistake Abby Dog made, as if my failures were public spectacle. The breeder's words loomed largest of all, though, their venom cutting deeper than I wanted to admit. I had chosen Abby Dog carefully, envisioning her as the farm's future, a dog who would one day bear pups to guard the farmstead. Instead, she was the source of endless chaos, reflecting my self-doubt.

I looked out over the disarray: Abby Dog, now sitting obediently in her exaggerated *I'm-a-good-girl-now* pose, chickens cautiously returning to peck at the ground, and Toby Dog standing like a sentinel, his watchful gaze full of judgment. The weight of it all settled heavily on my shoulders.

But even as doubt clawed at me, another voice, quieter but resolute, rose within. I couldn't give up on Abby Dog. Not again. Not like I had before.

When I first reached out to Bertha about adding a second livestock guardian dog to the farm, I expected hesitation. Most Maremma breeders tend to be protective of their dogs, carefully vetting potential buyers. But Bertha was excited to work with me from the start.

"You're a social media star," she wrote in one of her early emails. "It would be amazing to see one of my puppies featured on your farm. My husband and I love your videos!"

Bertha lived in Nevada, nearly three thousand miles from Vermont, and ran a large breeding operation. Her emails overflowed with updates about the puppies, photos of them tumbling through fields, and videos of their first interactions with livestock. She was eager to have me showcase Abby Dog's progress and even suggested joining a livestream to discuss her farm.

"The puppies are doing great! They are around 40 lbs. and so adorable!" she wrote. "I've been enjoying catching your show whenever I can. My husband has started watching it, too, and really enjoys it, which is funny because he's not the farmer guy type. The farm is my thing, not his, but he loves your show."

Bertha made it clear that temperament was her top priority, and she sent me regular updates as she assessed the puppies' personalities. She explained that some of the female pups had umbilical hernias and outlined the implications with impressive thoroughness: "Unfortunately, three of the four female pups have umbilical hernias. I had hoped they would close by now, and they may still be in time, but they haven't."

For breeding dogs, that's a complicated issue. It could be genetic, meaning it could be passed down to pups, or it could have been caused by the mom overcleaning the cords when they were born.

"There's no way to know for sure, but I strongly suspect it's the second cause."

Bertha reassured me that hernias could be easily repaired during spaying or neutering for dogs not intended for breeding. However, she cautioned that a separate surgery might be necessary for a breeding dog. Ultimately, I decided to choose the puppy that would become Abby Dog based on Bertha's evaluation of her temperament characteristics.

"She's a really pretty girl and very sweet. She's my favorite of all the girls. You and Toby are going to love her!"

Moving Abby Dog from Nevada to Vermont was like solving a logistical Rubik's Cube, except every side was colored *Inconvenience Beige*. Flying? Absolutely not. The thought of her riding in a cargo hold, scared and alone, made my stomach churn. Plus, she was probably too big at that point. Driving out myself wasn't much better. While the idea of a road trip with a new puppy sounded fun, between the farm, the animals, and everything else on my plate, a week-long cross-country road trip was out of the question. That's when I found Olivia.

Olivia ran a small, independent dog transport business, and after a few conversations, it was clear she had both the experience and temperament for the job. She promised regular updates through photos, videos, and texts, and she assured me the puppy would get the care and attention she needed during the cross-country journey.

"She'll ride in comfort," Olivia said on the phone. "And trust me, I've transported my fair share of lively pups. She's in good hands."

Olivia's first update came just hours after pickup: a photo of the puppy in the front seat, ears perked, looking like she'd just been recruited for an action movie. "She's curious about the open road," Olivia texted.

By day two, there was a video: the puppy bounding through Colorado snowbanks like she was auditioning for a Hallmark movie. By day three, she was making friends with a Golden Retriever at a rest stop, tail wagging like a metronome.

"She's a joy," Olivia wrote. "Very curious. Maybe too curious." That last bit came with a photo of the puppy eating a doggie ice cream cone and making a mess.

When Olivia's car finally pulled into my driveway, headlights cutting through the icy Vermont evening, I was outside before she could even park. The puppy bolted out, tail wagging, nose buried immediately in the nearest snowbank. And then, as if on cue, she bounded toward me, all floppy ears and wiggly excitement.

"Hi, puppy," I whispered, crouching down. The puppy responded by planting two paws on my chest and licking my face like she'd found her long-lost pack. At that moment, I decided that the name "Lady Abbington" would suit her well.

Toby Dog watched this chaos from a distance, his expression best described as "grumpy older sibling being told to share." Abby ran to him, tail wagging with Olympic fervor. Toby sniffed her once, then looked at me as if to say, *Seriously? This tornado?*

As Olivia drove off, I stood in the snow watching the puppy dart around, her energy lighting up the farm. Toby followed reluctantly as if he couldn't decide if she was a roommate or a nuisance. I didn't have all the answers yet. But watching her leap into another snowbank, I felt something I hadn't expected: relief. This was going to work.

Adjusting to life with Abby was a mix of comedy, chaos, and cautious optimism. From the moment she set all four flailing paws on the farm, it was clear that this would be no ordinary adjustment period. Abby had two speeds: full throttle and asleep. The quiet moments were rare, and when they came, I often wondered what she was plotting for when she woke up.

Farm life, with its steady rhythms, wasn't ready for Abby. Toby Dog knew the script. His days were marked by patrols, dignified naps, and stoic cameos in my YouTube videos—an old pro. Abby, by contrast, approached the farm like a theme park built just for her. Chickens? Flapping, honking chew toys. Ducks? Potential tag partners. The barn? A playground begging for chaos.

Her first full day on the farm was a circus. Morning chores were punctuated by Abby dashing through the barn, skidding on the icy patches, and sending a bucket of feed flying. Chickens scattered, squawking in protest, while Toby followed her movements with the weary gaze of someone who'd seen it all before. Every cluck, honk, or flutter of wings ignited a spark in Abby. She lunged, she barked, she pounced.

"Abby! *No*!" I shouted for what felt like the hundredth time that morning. She froze mid-pounce, turning to me with her head tilted, her expression equal parts confusion and defiance. I crouched down, my voice softer now.

"Abby, come here." Her tail wagged tentatively, and she trotted over, nose sniffing the treat in my hand. "Good girl," I murmured, scratching behind her ears.

Toby stood a few feet away, watching with what I can only describe as skepticism. If Abby's energy was a storm, Toby was the eye of it. His calm demeanor balanced her out—or at least, that was my hope. I needed him to show her the ropes, to teach her what it meant to be a livestock guardian dog. Whether he was up for the challenge, though, was another question entirely.

By the end of that first week, the honeymoon phase had completely worn off. Abby's boundless energy wasn't just some quirk. It was a problem. She treated every chicken like a squeaky toy and every goose like a new friend to chase. My voice was hoarse. My patience was thin.

One afternoon, after another chicken-chasing incident, I sat on the barn steps with Toby. Abby was in her designated puppy area, barking her displeasure at being separated from the action.

"I don't know, buddy," I said, scratching at his thick fur. "Maybe this was a mistake."

Toby let out a low huff, his gaze fixed on Abby as she pawed at the gate. He didn't offer any answers (he never did) but something about his steady presence kept me from spiraling further into self-doubt. Later that afternoon, she figured out how to scale fences and escape.

As the days turned into weeks, my optimism began to falter further. Training sessions felt like a constant two steps forward, three steps back. Abby was eager to please, but her instincts to chase and play overpowered her desire to listen. The sight of a duck waddling across the yard or a chicken flapping its wings was simply too much for her to resist. And while she didn't mean harm, the stress it caused the animals and me was undeniable.

It didn't help that everything was happening under the watchful eye of the internet. Every training session, every mishap, every moment of doubt felt magnified by the presence of a camera. My YouTube audience, once a source of encouragement, now felt like an audience for which I was performing. Comments rolled in after every video: advice, critiques, and, occasionally, outright criticism.

"Why would you bring a puppy onto the farm if you don't know how to train her?" one viewer wrote.

"Doesn't look like she has guardian instincts," someone wrote.

"Big mistake," added another. *"Not every dog is cut out for this."*

I tried to shrug off the negative comments, but they lingered in the back of my mind, feeding into the insecurities I already carried. Training Abby felt like a public spectacle, a mix of slapstick and tragicomedy. And every setback played out before an invisible jury.

At night, as the farm settled into its usual quiet rhythm, I often replayed the day's events in my mind. Inevitably, those reflections led me back to Barkley.

Barkley was the dog my parents gave me when I was twelve. He was a floppy-eared beagle with a boundless enthusiasm for life and no interest in rules. I had begged my parents to get him and sold it as a "responsibility-building exercise." But I hadn't been ready for a dog like Barkley. He chewed everything, barked incessantly, got aggressive around food, and refused to stay in our yard if he wasn't on a leash. After months of failed attempts to train him, my mother rightly decided it was time to rehome him. I hadn't put up much of a fight.

Even now, years later, the memory of Barkley's floppy ears and wagging tail haunted me. I told myself I'd done the best I could and that I'd been just a kid. But deep down, I'd always wondered if I could've done more. If I had given up too easily. To this day, it remains one of my biggest regrets in life.

Abby Dog wasn't Barkley. I was older now and more experienced. But the parallels were hard to ignore. Her energy, her stubbornness, her endless curiosity. It all felt painfully familiar. And the fear of failing her, of failing myself, loomed large.

"You're not going anywhere," I whispered one night, sitting with Abby on the barn floor after another exhausting day. She rested her head on my lap, her wide brown eyes gazing up at me with that same unshakable hopefulness. "I'll figure this out. I have to."

The next morning, I woke up determined to take a new approach.

But then Blanche happened.

Blanche was a white Leghorn chicken, one of the elder stateswomen of the flock, and I found her one snowy morning in a heap of feathers outside the barn. Abby stood a few feet away, her ears pinned back, tail wagging cautiously, an unsettling combination of guilt and misplaced pride. Toby wore his patented *I knew this would happen, but nobody listens to me* look.

I crouched down and examined Blanche's lifeless body. She hadn't just been killed. She'd been played with to death. I could see the marks from Abby's teeth. The places where she'd tugged and pawed, thinking it was all a game. My stomach sank. This wasn't just a setback. It was a Hindenburg.

"Abby," I croaked. She crept toward me, tail wagging nervously as if hoping for reassurance. I wanted to yell, to scold, but I knew it would do no good. She didn't understand what she'd done—at least, not entirely. And if I was being honest with myself, this wasn't her fault. The blame was entirely mine. Abby had snuck out of her pen and gained access to the chickens well before she was ready to be left alone. Blanche was the consequence of that mistake.

As I sat in the barn with Abby curled at my feet that evening, I thought about what to do next. Giving up wasn't an option. Not this time. If I failed her, it would haunt me. Just like Barkley.

The next day, I called a trainer specializing in livestock guardian dogs. I explained Abby's chaotic energy, Blanche's untimely demise, and my inability to fix the situation. The trainer listened quietly before smacking me with the hard truth.

"You're treating Abby like she's Toby," she said. "She's not Toby. She's Abby. You need to meet her where she is. Supervise everything. Structure her training. Constant consistency. And most importantly, be patient."

Her words stung, mostly because she was right. There was no shortcut. I needed to rebuild the entire process from scratch. Abby stayed on a leash or in a secure pen and was never left unsupervised. I created a small enclosure near the chicken coop so she could watch the flock without wreaking havoc. I also started using an e-collar with a vibrate function. Slowly, I began reintroducing her to the birds.

I also focused on bonding with her outside of work. We went for long walks, played tug-of-war to burn off her energy, and worked on basic commands. The more time I spent with her, the more I understood her quirks and the weirdness that was her brain.

Sharing the story of Blanche's death on YouTube was difficult. Knowing the harsh criticism, I debated whether to address it all. But I'd built my platform on honesty, and hiding the truth didn't sit right with me. One of the things I've learned about content creation over the years is that vulnerability is essential. If you can't get comfortable with putting yourself out there, you shouldn't make videos about your life. You've got to get comfortable sharing your failures, or people will quickly see through the charade.

So, I started regularly posting videos about Abby's struggles and failures. I walked my audience through what had happened, what I'd learned, and how I planned to move forward with Abby.

The comments ran the entire emotional spectrum: supportive, critical, brutally honest, and borderline unhinged. It was like hosting a town hall where everyone thought they were the expert.

And while I tried to take the helpful advice to heart, the criticism lingered in the back of my mind, rubbing against my insecurities like sandpaper. Some viewers were supportive, sharing stories about their struggles with high-energy dogs. Others weren't so kind, offering unsolicited advice.

"This dog isn't cut out for guarding."

"You're setting yourself up for more failure."

"Send her back before you lose more animals."

And then there was Bertha.

When Bertha's email landed in my inbox, I knew I was in trouble. "I do not understand why you are posting videos that paint Abby and me in a bad light after all I did for you," she wrote, her tone vacillating between hurt and fury. She detailed how she'd defended me against criticism from others, held onto Abby longer than planned, and promoted my videos across her social media. "And this is the result?"

I stared at the email, feeling a headache creep up the back of my skull. I hadn't set out to criticize Bertha. If anything, I had bent over backward to avoid directly assigning blame. But it didn't matter. Her email made it abundantly clear that my attempts to take full responsibility for Abby's struggles had landed about as gracefully as a goat on roller skates.

Soon, Bertha began publishing full-scale literary assaults that felt like equal parts impassioned essays and unhinged rants. In her initial posts, she accused me of mishandling Abby, being grossly inexperienced, and jeopardizing the future of livestock guardian dogs everywhere. "When a man who is held up as a 'Maremma expert' by over a half a million people tells the world that his NINE MONTH OLD PUPPY is a 'disappointment', a 'bad dog', and a 'failure'...that is a tragedy," she wrote. "Innocent dogs will die.... All for clicks and money."

Over the next month, Bertha unleashed what felt like the unabridged version of *War and Peace*, livestock edition. By my count, she churned out at least twenty thousand words dissecting every aspect of my videos, my training methods, and, bizarrely, the layout of my farm. Her posts had titles like "The Decline of Integrity in LGD Training" and "The Danger of Misinformation: One YouTuber's Recklessness." They were littered with lines like, "It's clear he doesn't understand LGDs.... Or worse, he doesn't care."

Some of her claims were almost comical in their hyperbole. In one post, she suggested that I'd been influenced by "wicked fairies whispering in my ear" to turn against her, as though my criticisms were part of some grand conspiracy. In another, she accused me of being "corrupted by peer pressure" from an imagined cabal of rival breeders. Without getting into all the specifics, Bertha had a lot of drama with other Maremma breeders.

(And I know a lot of the language I'm quoting here may seem bananas and exaggerated. But I went back through old emails, screenshots, and blog posts to quote accurately for this book.)

And yet, despite the melodrama, her posts hit hard. The relentless nature of her writing felt like a battering ram against my confidence. She didn't just accuse me of mishandling Abby. She painted me as a reckless amateur whose carelessness directly threatened dogs everywhere. I could laugh at the "wicked fairies" line, but other parts stayed with me, gnawing at the edges of my self-worth.

Yet as easy as it was to paint Bertha as the villain, part of me empathized deeply with her reaction. Being criticized publicly is never easy, but it's especially harsh when you're not used to the relentless gaze of the internet. Over the years, I'd built up a somewhat thick skin, developing calluses from the constant barrage of online critique. Bertha didn't have that armor. For her, each comment, each critique, must have felt intensely raw, a direct attack on her life's work, passion, and pride. Even though she had originally courted the positive attention, she wasn't prepared for thousands of strangers dissecting her decisions or second-guessing her intentions. Behind the bitter posts and dramatic accusations, I saw someone overwhelmed by the spotlight—or, in this case, more like the flashlight—flailing defensively against pressures she'd never signed up for and certainly hadn't anticipated. While her reaction stung deeply, I recognized the vulnerability beneath her outrage: a person simply unprepared for what happens when private disappointments spill into public view.

"It's true," said Allison. "I think Bertha's behavior is totally unhinged."

"Thank you," I said, relieved to feel seen.

"But," she added, not looking at me, "you do know there's nothing you enjoy more than a good fight."

"What's that supposed to mean?" I said, half-defensive.

"You're the type of person who is not afraid to kick a hornet's nest. In fact, you have fun doing it. It's almost like you get a charge out of finding out what will happen next. Until you start getting stung a few times. And then you're like, 'I don't like this. This hurts.' But you would rather be stung by a swarm of angry hornets than let one fly by in peace."

She wasn't wrong.

Bertha's words lingered as I tried to focus on work and push through. Her posts became a dark cloud hanging over everything I did. Every training session, every video I edited, every moment of doubt felt amplified by her accusations. It wasn't just that she criticized my methods. She seemed to think I was *fundamentally unworthy* of being Abby's owner.

What stung most was the hypocrisy. This was the same breeder who had once called me a "great advocate for the breed" and celebrated my decision to document Abby's training journey. Now, she was turning that very openness into a weapon used against me. She had no issue with my platform when it served her, but I became the villain the moment it didn't.

But then a different kind of email started showing up in my inbox. Because of the public nature of the feud, other folks who had negative experiences with Bertha started to reach out to me. These emails were supportive, sharing stories of similar behavior they'd experienced from her. One email stood out. A customer claimed Bertha had blocked them after they raised concerns about a "defective" dog and then launched a smear campaign against them on social media. Suddenly, I realized I wasn't alone in dealing with her wrath.

When I publicly addressed the feud in a video entitled "Spilling Tea on the Abby Drama," the damage was already done. "Because I'd gone publicly out on the record praising her, I also felt like it was important for my audience to correct the record and say, 'Hey, look, my feelings have changed here, and here's why,'" I explained. "I didn't feel like it would be an attack to give an honest update."

The video wasn't about clapping back or scoring points.

Well, maybe it was a little bit about clapping back.

Bertha's behavior wasn't my responsibility, but my response was. I couldn't control her blog posts, her dramatic accusations, or her apparent belief that I'd been possessed by "wicked fairies." I could control how I chose to move forward.

Abby's transformation didn't come with a eureka moment or a magical "aha!" It came with a hundred tiny, stubborn steps forward and a few spectacular stumbles back. Over time, we found a rhythm. Not flawless, not Instagram-worthy, but ours.

The e-collar was my lifeline. Not a punishment device but a glorified "hey, knock it off" button. A quick vibration here, a tone there, just enough to break her focus when her chicken-chasing instincts kicked in. Abby wasn't thrilled about it, but it worked. Little by little, she grasped that the chickens weren't squeaky toys on legs.

Abby's designated pen near the chicken coop was the other critical factor in her improvement. In the beginning, she hated it. She barked, pawed at the gate, and threw what could only be described as canine tantrums. But over time, the pen became her classroom. She'd sit and watch the birds, her head tilting as she tracked their movements. The energy that had once driven her to chase was still there, but now it was tempered by a growing sense of curiosity. She started to understand that the chickens weren't playthings. They were hers to protect.

Of course, she had help. Toby Dog, ever the stoic mentor, stepped into the role of unwitting teacher. He didn't bark at her or try to put her in her place. Instead, he patrolled with the kind of quiet confidence that said, *This is how it's done. Watch and learn.* Abby, whether consciously or not, started to mimic him. If Toby wasn't chasing the chickens, why should she?

The moment that stood out to me came one late afternoon while I was filling the water troughs. I heard a commotion from the chicken coop and immediately felt my stomach drop. Abby had managed to sneak in while I was distracted, and my first thought was, *Here we go again.* But when I got to the coop, I saw something unexpected. Abby was standing still, her head low, her tail wagging cautiously. The chickens, flustered but unharmed, were all gathered in the corner. And there was Abby, standing in the middle of it all, not chasing, not pouncing, just watching.

It wasn't perfect, but it was a start.

Abby's energy never disappeared. She still tore across the fields like a toddler who'd just discovered sugar, her enthusiasm both exhausting and oddly inspiring. But what had once been a liability started to feel like an asset. Some nights, I'd go out to check on the farm, and it would bring me joy to see the dogs in action. Together, they were like an odd-couple cop duo: Toby, the no-nonsense veteran, and Abby, the eager rookie with too much enthusiasm but a heart of gold. She was the definition of chaotic good.

One evening, I watched Abby and Toby patrolling the pasture together. Toby moved with his usual steady precision, the seasoned professional with nothing to prove. Abby, meanwhile, bounded beside him like an overenthusiastic intern, her steps quick, her tail wagging furiously at the sheer joy of being out in the fields. She wasn't just the chaotic puppy I'd brought home anymore. She belonged here, quirks and all.

Abby wasn't the dog I thought I wanted. She wasn't a Toby Dog clone or the mythical "perfect" livestock guardian dog I'd envisioned. But she didn't need to be. Abby was herself. Wild, determined, fiercely loyal, and full of a spirit that could exhaust you and inspire you in equal measure.

And as it turned out, she was exactly what ~~the farm~~ I needed.

CHAPTER 15

THE LEGEND OF MOLLY MURDER MITTENS

"Molly?" I called, my voice sharper than I intended. My breath puffed into the crisp morning air, but there was no answering sound. The only response was the barn door creaking lazily in the wind.

"Molly Murder Mittens, where are you?"

That morning carried the first signs of fall. It was the kind of morning that made you reach for a hoodie on your way out the door only to discard that very same hoodie once you started to get too sweaty during chores. A low mist hugged the pastures, blurring the edges of the landscape with trees turning orange, but the grass remaining stubbornly green.

On the porch, Pablo Barn Cat lounged on the railing like an indifferent king surveying his kingdom. Ginny Barn Cat was curled up on the freshly stacked firewood, stretching as if she'd been working all night instead of sleeping. Neither of them so much as twitched an ear at Molly's name. Of course not. They were barn cats, unfazed by the worries of humans or the vanishing acts of their own kind. *She'll come back*, their languid tails seemed to say.

"Where's your mom?" I asked Ginny, bending to scratch her behind the ears. She leaned into the touch but didn't offer any answers, only a lazy blink before returning to her feline version of a morning stretch.

Usually, Molly was the first to greet me. The moment I opened the mudroom door, she'd appear, rubbing against my legs and chirping impatiently for breakfast. While the other barn cats waited for me to finish chores, Molly was always there, shadowing my every move like a small gray sentinel. Her absence this morning felt strange. Out of step with the rhythm of the farm.

I kept busy with chores. Feeding birds and moving cattle. Typically, Molly would dart ahead of me, weaving under gates and perching on fence posts as if overseeing the morning operations. Her sharp green eyes would track my every move, judgmental and curious at the same time. Today, her usual haunts were empty: the woodpile, the tractor, even the spot under the barn eaves where she liked to sit and watch for mice.

I thought about the time she earned her regal title, "Murder Mittens." Two summers before, in the heat of July, she showed up on the porch with a mouse dangling from her jaws like a trophy. She dropped it at my feet, swiped a paw through the air as if to say, *This is how it's done*, then sauntered off, tail high. Pablo had glared at her for a full hour, and Ginny had sulked by the woodpile for the rest of the day. Molly didn't care. She never did.

Returning to the porch, I grabbed her stainless steel food bowl and filled it with kibble. I shook it, the sound echoing into the yard like an open invitation. "Molly!" I called. "Breakfast!"

The rattling usually brought her running. She'd leap onto the porch indignantly to remind me that feeding her should have been my first task. But today, the bowl sat untouched, the kibble glinting in the early light.

The breeze picked up, carrying the faint rustle of the maples lining the property. Pablo stretched and leaped down from his perch, trotting toward the barn as if he'd grown bored with waiting. Ginny followed, her tail flicking in a way that suggested she wasn't concerned in the slightest.

But I was.

I stood on the porch for a long moment, scanning the yard. The mist was beginning to burn off, revealing the familiar outlines of the barn, the hoop coop, and the fields stretching beyond. The farm carried on as usual. The cattle lowed softly in the distance, and the ducks quacked as they looked for bugs in the permaculture orchard. But without Molly Murder Mittens, the day felt unbalanced, like a chair missing a leg.

Setting the bowl of kibble back on the porch, I sighed, the worry settling deeper into my chest. I told myself Molly was probably off chasing a mouse or staking her claim somewhere new. According to the stories, barn cats had a way of disappearing and reappearing on their own terms.

Still, I couldn't help but feel the quiet gnaw at me.

Molly and Ginny had come to the farm in early June 2021. Logan, a friend of Allison's, had found Molly in her barn that spring. She was heavily pregnant and feral (Molly, not Logan) a mother cat navigating tough odds. Molly gave birth in a pile of hay, raising six kittens with the ferocity of a mother who'd seen hard times.

By June, Logan had found homes for most of the kittens, but the mom and one remaining kitten still needed homes. When Allison told me about the situation, it seemed like fate. With Pablo stretched thin patrolling the farm and Lil Barn Cat now permanently living indoors after her accident, we needed more barn cats to keep the rodents in check. The idea of bringing on a mother-daughter duo just made sense.

Logan's daughter had already named the kitten Ginny. I thought Molly was the perfect name for the mother. Molly and Ginny. It had the right ring to it, like characters from a story by You-Know-Who.

On the day they arrived, Logan opened up the hatch of her car with the pair in a borrowed cat carrier. I could hear their soft meows even before she stepped out. Molly's sharp green eyes glared back at me from inside the carrier, her body hunched protectively around Ginny. The kitten, a smaller version of her mother, peered out from behind Molly's tail with wide, nervous eyes.

"They're good cats," Logan said as she handed me the carrier. "The mom is a top-notch hunter. Just give her a little time to adjust."

The room I'd prepared for them was on the second floor of the old three-story barn. It had likely been a chicken coop in decades past. Still, with some repairs and reinforcements, it became the perfect space for acclimating barn cats. I'd added chicken wire over the windows, lined the walls with hardware cloth to keep them safe, and filled the space with straw, food, water, and a litter box.

"Here you go," I murmured, carrying them inside and setting down their transport.

Molly was the first to emerge. She stepped out with the precision of a predator, her tail flicking as she sniffed the air. Ginny stayed hidden, her tiny body jammed against the back of the cat carrier. But with some gentle coaxing, she followed, sticking close to her mother as they explored our strange barn.

Over the next few days, Molly and Ginny settled in. Their personalities quickly became apparent. Molly was sharp and no-nonsense, a cat who carried herself with the confidence of a creature who had survived on her own. Even back then, Ginny was an aggressively friendly kitten who might have been born lacking a few brain cells.

After a few weeks, it was time to let them out into the barnyard. I opened the door to their room and stepped aside, watching Molly lead the way. She was ready to take ownership of our farm. Ginny followed a few steps behind, her tail twitching with nervous excitement.

It didn't take long for Molly to stake her claim. Within days, I started finding signs of her work—mice left on the porch, neatly arranged as if to remind me she was already earning her keep. Ginny stayed close to her mother, learning the ways of the barn with wide-eyed attentiveness.

By the end of the summer, they had claimed their place on the farm. Molly patrolled the yard with quiet authority, her green eyes missing nothing. Ginny's kittenish clumsiness had started to give way to clumsy adulthood, though her friendly disposition still made her the kind of cat who would rub up against a cow just to see what would happen.

Molly wasn't just another barn cat. She was a presence, a hunter, and a mother raising her daughter to do the same. And while it would be weeks before I'd see her soft side, I knew we'd made the right choice bringing her to the farm.

Molly Murder Mittens had a way of moving through the farm that made you notice her only when she wanted to be seen. In the mornings, I'd catch her perched on a fence post, her tail curled like a plume, watching the barnyard with a sharp, knowing gaze. By the time I'd finished my chores, she'd be gone, slipping into the shadows of the barn or stalking along the edge of the bird yard, invisible except for the occasional flicker of green eyes through the grass.

Ginny, by contrast, didn't seem to have a subtle bone in her body. She tore through the yard like chaos on four paws, her intrepid energy consistently outstripping her coordination. Once, I saw her leap at a patch of dandelions with the conviction of a lioness on the Serengeti. When the stems refused to cooperate, she flipped herself backward into the dirt, landing in an ungraceful heap. Above her, Molly watched from a fence post, her tail flicking slowly—a queen observing a jester—but she didn't twitch a whisker to intervene.

One afternoon, I found Molly sprawled on the barn's stone threshold, basking in the golden light of late summer. Ginny lay a few feet away, cleaning her paws with the over-exaggerated focus of a student trying to impress their teacher. I crouched nearby, quietly watching as Molly's ears flicked toward the faint rustle of leaves beyond the bird yard. She wasn't lounging. She was listening.

The rustling continued. Molly rose with practiced ease. Her movements were always fluid and deliberate. And silent. Ginny tried to follow her mother's lead, crouching low, but her twitching tail betrayed her kittenish excitement.

The source of the sound revealed itself. A vole creeping along the edge of the barnyard. Molly stalked forward, silent as smoke. Ginny stayed behind, her paws kneading the dirt as if she could barely contain herself.

Molly struck fast and clean. One sharp swipe of her paw, and it was over. She carried the vole a few feet, then placed it neatly on the ground, nudging it toward Ginny. Ginny pounced with unearned confidence, tumbling into the grass and losing her grip. Molly sighed, if a cat could sigh, and picked the vole up again, carrying it toward the barn without a backward glance.

Ginny sat in the grass momentarily, her tail twitching in frustration, before trailing after her mother. The lesson, I supposed, wasn't just about the hunt. It was about patience, precision, and knowing when to act.

I often thought of Lil Barn Cat, who had learned that lesson the hard way. She had been our tiniest, scrappiest mouser until the day she darted across the road at the wrong moment. Lil was a reminder that everything out here is temporary, a fleeting balance of life and loss. Barn cats, especially, live in that delicate in-between. They're resourceful and independent, but the freedom that lets them thrive makes them vulnerable.

The risks were always there, woven into the fabric of farm life like threads you couldn't entirely untangle. Sometimes, I'd catch sight of a hawk, its silhouette cutting lazy circles high above the fields, its shadow gliding silently over the barn. Other times, the coyotes would announce themselves in eerie harmony, their howls rolling through the tree line and carried on the wind.

On the farm, the livestock guardian dogs are the front line of defense, and their presence is a clear deterrent for predators like coyotes. But the barn cats are the unsung heroes of this delicate balance. Predators like coyotes, foxes, and bobcats survive largely on rodents in our part of the world. By keeping the rodent population in check, the barn cats indirectly discourage larger predators from venturing too close to the barnyard.

Dogs bark, pee on trees, charge the fence line, and stand their ground. On the other hand, the cats do their work in the shadows, slipping unnoticed between hay bales and fence posts. Their contribution is invisible until it isn't—a rodent-free barn, a silent field, a moment of peace held in place by their unseen efforts.

That night, as I closed up the barn, I saw Molly and Ginny perched on a straw bale, their silhouettes outlined in the last light of the day slipping through the cracks in the barn plank wall. Molly sat upright, regal, and composed. Ginny was sprawled beside her, her tail flicking in a playful back-and-forth. The two of them seemed so perfectly at ease, so woven into the fabric of the farm, that it was hard to imagine them as anything but permanent fixtures.

But nothing on the farm is permanent.

The first morning Molly didn't appear, I brushed it off as a quirk of barn-cat independence. By the second, it felt like an alarm bell I couldn't ignore.

It started the same as any other day. I stepped out onto the porch, the air carrying that early September mix of crispness and the faint warmth of lingering summer. Pablo was perched on the railing, casually swiping at a fly, and Ginny was curled in the straw by the barn door. But there was no sign of Molly.

"Molly?" I called, shaking the battered kibble dish in my hand. Ginny stretched and let out a small chirp, looking at me like she was trying to reassure me. But when I crouched down to scratch her ears, her eyes flicked toward the fields, scanning the distance.

I told myself not to worry. Barn cats vanish all the time, off chasing some distant rustle in the woods or staking their claim on a forgotten corner of the farm. They disappear, and then they come back, slipping onto the porch with an air of casual indifference as if they'd never been gone.

But by the third day, Molly still hadn't returned.

Allison stepped onto the porch, tightening her sweater against the morning chill. "Still no Molly?" she asked softly. I shook my head. Her sigh echoed my worry. "She's tough," she added reassuringly. "She'll be back."

I nodded, but didn't say anything. Allison knew that look. I was spiraling.

"You've done everything you can," she said, sitting down beside me.

"Except find her."

She let that sit for a second. "You know you do this, right?"

"Do what?"

"Convince yourself that if you just work harder, you can force the outcome."

I obsessively searched Molly's usual haunts, including the barn, the woodpile, the shadowed corners near the tractor shed. I walked the edge of the bird yard and checked beneath the hay bales, calling her name even though I knew she wouldn't come running. Molly wasn't the kind of cat you summoned. She appeared when she wanted to.

Ginny trailed behind me as I searched, her tail flicking nervously. She kept glancing over her shoulder as if she expected her mother to materialize from the grass and take over the lesson she wasn't ready to lead.

By the end of that day, I couldn't ignore the knot of worry tightening in my chest.

I widened my search the following day, driving slowly down the dirt road that borders the farm. With the windows down, I called her name, scanning the ditches for any movement. Every shadow made my heart catch, every flash of gray fur turned out to be a squirrel or a clump of weeds.

I posted on the local Front Porch Forum, describing her in as much detail as I could: *Small gray tabby with green eyes. Answers to Molly or doesn't answer at all, depending on her mood.*

(For non-Vermonters, Front Porch Forum is like if Craigslist had a baby with Nextdoor, and then that baby was adopted by your local Facebook community group but somehow still manages to be pleasantly civil. It's where you go to find lost cats, sell junk, debate the ethics of leaf blowers, and watch your neighbor passive-aggressively complain about someone's dog, all while reading it in a format that feels like your inbox time-traveled back to the AOL era. It's quaint, chaotic, and weirdly essential.)

The replies trickled in slowly. Well-meaning neighbors suggesting cats they'd seen, none of them Molly.

The farm carried on, indifferent to my growing worry. The chickens scratched, the geese honked their disapproval when I took too long with their water, and the cattle grazed as lazily as ever. But everything felt muted, as though the farm itself had noticed a missing pair of murder mittens.

Each evening, I lingered on the porch longer than necessary, staring at the horizon until the light faded. I continued to leave a third dish of food out for Molly, even if Pablo or Ginny would probably eat it overnight.

One afternoon, about a week into my search, I thought I saw her. A gray shape darted through the tall grass at the edge of the tree line, and my heart jumped.

"Molly!" I called, moving quickly toward the spot.

The shape froze, and for a moment, I was sure it was her. But then it moved again. A rabbit, startled by my voice, bounded deeper into the woods.

When I returned to the barn, my chest felt heavier than before.

The second week without Molly began the same way as the first. A dull ache I couldn't quite locate. It wasn't just missing her sharp green eyes or the soft, insistent chirps she used to remind me who was really in charge. It was the way the quiet felt louder, like the spaces she once filled were now actively conspiring to remind me she wasn't there.

I kept replaying little moments I hadn't realized were so deeply etched into my day-to-day. How Madame Murder Mittens would perch on the porch at sunrise, her tail flicking in clear judgment: *What took you so long?* She'd leap onto a straw bale as I stacked them, her green-eyed gaze making it clear she could do the job better if only she had opposable thumbs. Or her favorite spot by the bird yard fence, sitting like a tiny furry warden overseeing the chaos of honking geese and clucking hens with what I can only describe as feline disdain.

And then there was that thunderstorm. One of those relentless summer tempests seemed to pound the farm from all sides. I found her curled up in the barn, her gray fur melting into the shadows like she was part of the architecture. The wind howled, the rain lashed against the roof, but Molly? She just blinked at me, utterly unbothered. I'd sat beside her, letting the storm rage while she radiated a calm so solid it might as well have been bolted to the floorboards.

Now, every corner of the farm seemed to magnify her absence. The afternoons were the worst. That was her prime patrol time when she moved through the barnyard like a tiny general inspecting the troops. She'd pounce on grasshoppers with the precision of a laser-guided missile, then vanish under the tractor in hot pursuit of a mouse.

Ginny tried to pick up the slack, but she continued to be more chaotic than competent. She'd barrel through the barnyard, leaping after bugs with all the grace of a tipped wheelbarrow, and bounce up the straw bales as if she'd invented parkour. But even Ginny seemed off her game, stopping occasionally to glance toward the fields. It was like she was waiting for Molly to stride back in, ready to resume her role as the undisputed queen of the farm.

It was the kind of absence that made you realize how much you take the small, ordinary moments for granted until they're gone.

Then, one afternoon, I got an email.

"I think I saw your cat," a neighbor wrote. "A gray tabby, out by the lake."

My heart leaped. The lake was only a mile and a half away. The thought of my cat wandering that far sent me a new wave of worry, but it was quickly overtaken by the hope that this might finally be the break I'd been waiting for.

I grabbed a small container of kibble and a flashlight, even though it was still light out, and headed for the car.

The road was quiet when I arrived, the faint peeps of frogs filling the warm dusk air. I parked near the lake's beach parking.

"Molly!" I called, stepping out of the truck and shaking the kibble dish. The sound rang out in the stillness, but no gray shape emerged from the undergrowth.

I walked the water's edge, peering between the trees and calling her name. At one point, I thought I heard a rustle in the brush, and my heart raced. I followed the sound, stepping carefully over the uneven ground, and froze when I spotted movement near a cluster of wildflowers.

It was a cat.

"Molly Wobbles?" I whispered, crouching low and holding out the kibble dish.

The cat turned, its yellow eyes catching the light. For a moment, my breath caught. It looked so much like her: the same white neck, sleek gray coat, and sharp, intelligent gaze.

But then it stepped fully into view, and the illusion shattered. This cat was male, his frame broader, his fur darker and less refined. He stared at me momentarily, then turned and slinked back into the trees without a sound. I stood there, the kibble dish still in my hand, and felt the weight of the hope I'd carried here dissolve into the late afternoon air.

When I returned to the truck, the light was beginning to fade, and the world felt impossibly quiet. I sat for a moment, staring out at the rows of trees, and thought about how easily I'd let myself believe it could be Molly.

The drive back to the farm was slower than it needed to be. When I pulled into the driveway, the familiar sights of the barnyard came into view. Pablo perched on the porch railing. Ginny darted through the grass. But the emptiness where Molly should have been felt sharper than ever.

I lingered by the barn that evening, the kibble dish still in my hand, and stared out toward the pasture. Somewhere, Molly was out there. She had to be.

It took me more than a month to admit what I'd known in my gut for weeks: Molly wasn't coming back.

That fact finally hit hardest when I sat down to record the video. For weeks, I'd avoided it, clinging to the faint hope that she'd strut onto the porch one morning, her tail flicking as if to say, *What's all the fuss?* But as the days stretched into weeks, that hope became harder to carry.

Sitting in front of the camera with the barnyard behind me, I started talking. The words came slowly, halting and uncertain, until I finally said it: "I don't think Molly is coming back."

Hearing myself say it out loud made the loss feel real in a way nothing else had. I stumbled through the rest of the video, trying to explain what she meant to the farm, to me, knowing full well I'd never find the right words.

For months after, the questions kept coming. Every TikTok video, every YouTube comment section, there they were: "Did you ever find Molly Murder Mittens?" "What happened to Molly?" "I hope Molly's okay."

They meant well. They were kind, even. But every question felt like a fresh knife-wound reminder of her absence, a wound I thought was starting to heal but never fully could. I wanted to tell them how often I still scanned the fields out of habit, my heart catching at every shadow or flash of gray. I wanted to explain how it hurt to keep saying, *No, we haven't found her.*

But I couldn't keep unraveling it, not without tearing open the grief all over again. Even now, as I write this, tears stream down my face while Lil Barn Cat glares at me from her perch in my office, as if to say, *Get a grip, bruh.*

Before the farm, my life was tidy, organized, and distant. I kept people at arm's length, guarded against anything that might tug too hard at my heartstrings. I had no pets because they would make life complicated and messy. Back then, there wasn't much to lose, but there wasn't much to hold onto. My days were often lonely and empty from a lack of meaningful connection.

The farm turned all that nonsense upside down. It pulls you into its rhythm and forces you to invest in every little thing: the animals, the land, the routine. You fall for the creatures that surround you, be it a goose with a bad attitude or a barn cat with piercing green eyes, and when you lose them, heartbreak is inevitable.

And yet, that heartbreak is worth it.

I like to believe that Molly's story didn't end with her disappearance. In my mind, she wandered into the life of someone who needed her even more than I did. Perhaps an elderly widow living alone saw a gray tabby at her back door one evening and welcomed her in to bring some companionship and meaning to her life. I can picture Molly rolling onto her back, letting out her soft chirps, charming her way into a new home. I imagine her stretched out on a sunlit railing, tail flicking as she surveys her new kingdom.

I will never know what happened to Molly Murder Mittens. And that uncertainty will always sting a bit. But her life, however brief, mattered. It mattered in how she perched on fence posts, tail flicking with quiet authority. It mattered in the mice she left on the porch, her silent contributions to the farm's delicate balance. And it mattered in how she made me pause every morning, bending down to rub her exposed white belly while she purred with approval.

Grief doesn't fade. It transforms you. It weaves itself into the fabric of who you are, a reminder of the love that came before it. That lingering ache isn't proof of weakness or despair. It's the sign of a life filled with love. Loss only hurts because something mattered.

Molly isn't just a memory. She's part of this farm, stitched into the fabric of its story. I still feel her presence each day. In the dooryard. In the quiet corners of the barn. In the rhythm of this life she helped shape.

Molly Murder Mittens was more than a barn cat. She was a queen, a hunter, and a companion.

And now, she is a legend.

CHAPTER 16
NO ONE STANDS ALONE IN A STORM

The rain had been relentless all evening. Sheets of it pounding against the windows, overflowing the gutters, and turning the pastures into shallow lakes. A river had carved its way through our dooryard. For the second time in twelve months, Vermont was experiencing once-in-a-century flooding.

Allison was working the evening shift in the ER of the local hospital that night. She answered on the second ring, her voice low but steady.

"Hey. Is everything okay?"

"Yeah, we're fine," I said. "But listen. Don't try to come home tonight."

"What? Why?"

"Water's over the roads. There's no way in or out of Peacham right now. It's not safe. I already booked you a room at that motel near the hospital. It's under your name."

"You didn't have to do that."

"I know. But I also know you. You'd finish your shift and try to white-knuckle it back in the dark."

She sighed. "You're probably right."

"I'm definitely right."

She laughed, tired but genuine. "Alright. I'll stay. Text me if anything changes."

"I will. Love you."

"Love you too."

I had been out doing my rounds, checking on the animals and the outbuildings. A few of my neighbors' homes sat lower than mine, closer to the water. I checked in with them earlier, ensuring they had what they needed, but there was nothing anyone could *do* except wait it out and hope.

By the time I had stepped back inside the house, my clothes were damp with mist and spray, my soaked socks leaving wet prints across the floor. I was peeling off my jacket when the knock came.

Hard. Fast. Urgent.

I hesitated for half a second, just long enough to register that this wasn't a casual visit, before yanking the door open.

A local volunteer fire department member stood there, soaked to the skin, the reflective stripes on her fire department jacket glowing in the porch light.

"We need help," she said, breathless from either running or adrenaline. "One of Matt Kempton's sons is missing."

I stared at her for a beat, my brain catching up to the words.

"Which one?" I asked, as if that clarification would make the news any less distressing.

"I don't know."

A hard knot of worry settled in my gut. The Kemptons are a fixture in this town. A farming family through and through, the kind of people who would show up with a large excavator and help you dig trenches to plant hundreds of trees. Matty's sons, Amos, Will, and Dylan, weren't kids. They were men in their thirties. Experienced and capable. If one of them was missing, it wasn't because he got lost.

It was because something had gone wrong.

I put on my soaked raincoat while awkwardly trying to shove my damp feet back into boots I'd mistakenly thought I'd worn for the last time tonight.

"Where are they searching?"

"Down by the brook."

That made my stomach drop.

South Peacham Brook ran along the northern edge of my farm, about fifty feet below the pasture, separated by a steep embankment that dropped down to the road. I'd stood countless times at the pasture's edge, admiring the peaceful brook below. But that night it had transformed into a roaring and angry river.

I stepped out onto the porch, letting the rain hit me full force. The wind had picked up, driving the water sideways and soaking my pants before I even descended the steps.

We hurried across the yard, flashlight beams bouncing against the ground. We hopped into my side-by-side, the one with the winch and tow cable, and drove to the back corner of the pasture. I could barely make out the trees along the pasture's edge, their branches thrashing wildly.

We stopped at the pasture's northern edge, overlooking the steep slope that plunged fifty feet down to the road. Even in the dark, I could see how high the water had gotten. The brook had long since spilled over, claiming the pavement, dragging debris along in its current. A massive road culvert that was usually buried well below the ground was fully exposed.

Far below the point where we stood on the ridge, flashlights bobbed in the distance as other rescue folks searched along the submerged road. Tiny dots of light in the flood of darkness.

"Where am I least likely to screw things up?" I asked, aiming for humor but genuinely meaning it.

"Up here, along the embankment," she said. "We've got people down on the road, but we need eyes up here, too."

I nodded, adjusting my grip on the flashlight, and started walking. It was clear that there were search parties on the north side of the flooding. We would have to be the searchers on the south side.

The embankment's edge was slick with rain, mud, and the grass flattened from the downpour. My boots squelched with every step, sliding dangerously close to the drop-off. I could barely see ten feet in front of me. The flashlight's beam bounced off the rain, turning the darkness into a chaotic blur of water and shadows.

I stepped forward, sweeping the light in wide arcs, searching for anything: footprints, a shape, a sign of movement. My boots hit loose mud near the embankment's edge, and suddenly I slipped toward the water below.

My stomach lurched. I flailed, grabbing at the tall grass, my feet scrambling for purchase. For half a second, I thought I was going over, that I was about to plummet into the black, rushing water.

I grabbed a maple sapling jutting from the slope, knees sinking into mud, fingers gripping slippery leaves. My breaths came hard, fast.

I pushed myself back, careful this time, heart hammering in my chest. The embankment was a death trap. One wrong step, one patch of loose soil, and I'd be swept away.

I pressed on, slower now, more deliberate, scanning the flood below. The lights from the road looked impossibly far away. The sound of the water was deafening.

We searched like that for nearly a half hour, moving methodically, flashlights cutting through the dark, looking down on the water. To be more accurate, the firefighter searched; I mostly bumbled around and tried to stay out of the way.

Then, finally—

A crackle of static over a radio. A voice. Words that I couldn't quite hear over the wind and the rain. I turned, slipping slightly on the wet ground, straining to hear.

The firefighter beside me lowered her radio.

"They think they found him," she said.

A rush of relief hit me so fast, it was nauseating.

"Well, all's well that ends well." I declared.

I trudged back toward the farm, soaked to the bone, exhausted. I peeled off my dripping clothes, climbed into bed. Despite that night's adventure, everything was going to be OK.

Morning came sluggish and gray, the air thick with leftover rain and the smell of wet earth. The storm had finally moved on, but it had left its mark.

I pulled on my boots, still soggy from the night before, and stepped outside.

The farm was battered, but standing.

That was the first thing I checked.

While the chickens seemed a bit soggy, no animal is more joyous than a duck after a flood. The wide-eyed and restless cattle milled around in the main pasture, their hooves sinking into the mud. The pigs had it the worst. The flood had swallowed their yard, leaving them stranded on a tiny patch of high ground, an island in a sea of brown water. Pig Island! But they were healthy, at least, and not going anywhere. I'd deal with them later.

For now, I needed to check on my neighbors. I knew folks would be without power and internet, but we still had resources to offer at the farm. And while our tiny rural neighborhood looked bad, I wasn't expecting anything terrible.

My first stop was Frank's place. He was standing outside when I arrived.

"Hey, Frank! How did you and Patrice make it through the night?"

There was no small talk. No nod of acknowledgment, no *hell of a night*. Very un-Frank. He just stood there in the mud, hands shoved into his pockets, his face carved from stone.

"Dylan Kempton died last night."

It felt like a trap door opening in my chest.

"No," I said, shaking my head. "They said they *found* someone. I was out there last night. They found someone."

Frank exhaled through his nose and nodded. A slow, deliberate breath, like he was trying to keep something steady.

And just like that, I was crying.

The last time I had seen Dylan—like really seen him, not just a two-fingered wave when our trucks passed on a narrow dirt road—was over a game of cards. Dylan, Will, Matty, Frank, Chris, Ken. The usuals. He was quick with a joke, always. He had this way of grinning just before delivering a punchline, like he was letting himself enjoy it first before sharing it with you. The kind of guy who made a room warmer just by being in it.

But beneath the humor was a man who carried real weight on his shoulders. Dylan wasn't just a farmer. He was a steward of something bigger than himself, part of a family that had worked this land for generations. He grew up knowing what it meant to be responsible for something, to care for it, to put in the kind of work that doesn't come with applause. When a neighbor needed a hand, Dylan showed up. Not because he wanted recognition, not because he expected anything in return, but because that's just what he did. He was the kind of guy who would drop everything to help rescue a stuck tractor or brush hog an overgrown field. He never made a big deal out of it. He just rolled up his sleeves, got to work, and made it seem like no trouble at all.

He loved his family fiercely, and he had built one of his own with a wife, two young boys who were just starting to get old enough to follow him around the farm, to climb up on the tractor, to watch and learn. He had so much more life to live, many more lessons to pass down, many more hands of cards to play, jokes to tell, birds to hunt, and beers to crack open on a summer porch. And now, just like that, he was gone. It didn't make sense. A man like Dylan wasn't supposed to just *disappear*. The flood had taken him, ripped him away as if he were just another piece of debris. But he wasn't. He was *Dylan*. A father. A son. A friend. A neighbor. A man whose absence left a hole that would never quite close.

I wiped my face with the back of my hand, trying to get myself under control.

"How?"

Frank shifted, his boots making a sucking sound in the mud.

"He was out checking on the farms last night, on the road, when he got caught in the storm surge," he said.

I squeezed my eyes shut. I had been right above there. Standing on the embankment, scanning the flood with my flashlight, nearly slipping in myself. Calling into the dark, hoping for an answer.

And the whole time, Dylan was already gone.

I had gone to bed thinking we had done our part. That the worst was over. Now, I felt sick with the knowledge of how wrong I had been. I wanted to say something. Something useful. Something that mattered.

But all I could do was stand there, swallowing down the awful truth.

I left Frank's place in a daze. The road was a mess of mud and scattered debris, the ditches choked with broken branches, plastic bottles, and someone's overturned mailbox. The storm had taken so much.

Back at the farm, Alfred was already waiting, warming up his excavator. He had been using it the previous week to help with a farm project. The night before, I had moved it to higher ground when the flood waters started to rise. Now that decision felt like the only useful thing I'd done in the last twenty-four hours.

"This thing's gonna be critical," he said. "We can clear access, get people in and out."

I nodded, already thinking through the next steps.

After checking on all our immediate neighbors, I loaded up the side-by-side with supplies—gloves, tools, water—and headed out. The first priority was figuring out just how bad things were.

The roads were a mess. Some were completely gone, swallowed by the water. Others were covered in debris from trees, broken fences, and chunks of asphalt ripped up like loose fabric.

I sent the drone up. From the air, the damage was even worse than I imagined. Sections of the town were cut off, washed-out driveways leading to nowhere, uprooted trees everywhere. I fed the footage to the town, trying to help emergency and road crews figure out where to go first.

I took the side-by-side down back roads, stopping where I could.

Some folks were already digging out, sleeves rolled up, mud-streaked and determined. Others stood in shock, staring at what was left of their homes, barns, and lives.

"You need anything?" I asked one farmer neighbor, standing in the wreckage of his driveway, a tractor powering a generator that kept the farm's milk tanks cold.

He exhaled sharply, shaking his head. "No. We're fine."

"Well, we still have electricity and internet at my place if you guys need anything."

My offer of internet in this situation was about as welcome as offering vegan options at a pig roast. Furthermore, while most of my neighborly relationships were very good, things were a bit more tenuous on this specific farm. These folks were never my biggest fans, not quite sure what to make of the weirdo with the ducks and trees who had moved in down the road. But things had turned decidedly frosty after my highly publicized run-ins with the hound hunters. They frequently opened up their farm to the hound hunters.

"Well, if you do need anything, let me know," I offered with false cheerfulness. I continued my rounds.

By the time I got back to the farm, the neighborhood had been transformed yet again. The response to the flood was immediate, relentless, and entirely driven by the people of Peacham. There was no waiting for the government to step in, no hesitation. Within hours, neighbors were helping neighbors, clearing debris, ferrying supplies, and figuring out what needed to be done next. People showed up with whatever they had—excavators, chainsaws, dump trailers, bare hands—working side by side in the thick mud, lifting ruined furniture out of flooded homes, digging out driveways, and shoveling mountains of silt left behind by the water.

Allison had turned our massive wraparound porch into a makeshift internet café, complete with chargers carefully arranged along the railings and extension cords snaking neatly from the house. Handwritten notes taped to the columns provided Wi-Fi passwords with playful messages ("Wi-Fi password: duckduckbarncat - enjoy responsibly!"). Allison moved calmly between visitors, offering reassurances, gentle humor, and a sense of normalcy in the midst of chaos. Her thoughtful preparations transformed our porch into a gathering point of comfort and community.

And there was still more to do.

Two days after the flood, we turned our attention to a house that couldn't be saved. The two schoolteachers who lived there had barely escaped with their daughters and pets, fleeing as the water overtook their home. Now the fire marshal was coming. Once the house was condemned, everything inside would be lost for good. Twenty years of memories, the small, irreplaceable things that make a home, were still inside. Once the house was condemned, that was it. No going back inside.

So, in the morning, the town showed up. Dozens of people. Farmers, carpenters, doctors, lawyers, retirees. People who had their own flood damage to deal with but had set it aside because this was what mattered that day.

The only problem was *logistics*.

With the road washed out, no trucks or trailers could get near the house. The only way to move everything was by hand across the flooded South Peacham Brook, through the woods, up and over a steep bank where a caravan of ATVs and trailers waited to ferry everything out. From there, the route wound through pastures and back roads for nearly a mile before finally reaching solid ground, where trucks could load up and haul things to storage.

It was ridiculous. It was impossible. It was the only option we had. And so, we got to work.

At first, it was chaos. A dozen people were inside, hauling out dressers, mattresses, and stacks of books. Others outside, knee-deep in mud, forming a relay line to pass everything across the gulch. Out at the ATVs, drivers were loading as fast as they could. Coolers, lamps, framed pictures, a lifetime of belongings piled into the backs of trailers and strapped down with bungee cords. The machines roared to life, bouncing over rough trails and disappearing into the trees before returning empty and ready for another load.

And that's when things went a bit squirrelly.

The thing about disasters is that they make people frantic. But they also make people creative. In the chaos of the flood response—locals hauling soggy furniture, volunteers racing from Point A to Point

B—everyone was looking for ways to move faster, to do more. So when someone discovered a rough trail that cut a quarter mile off the main route, folks jumped on it. Efficiency, right? One less bend in the road, one more load delivered. The only problem? Nobody had checked with the farmers who owned that particular patch of land.

I knew the neighbors who owned the field. It was the same farm I had stopped at the day after the flood to check and see if they needed anything. The ones with the tractor-powered generator and no use for the internet. I was pleasantly surprised that they had opened their pasture to the volunteers. I wrongly assumed somebody had asked them for permission before we started making that part of the caravan route. But we all know what happens when you assume. It's a one-way trip to the donkey farm.

So there we were, using our newfound speedway, when I saw the tractor. Not just any tractor; this one had a tree in its front-end loader. The guy driving was a local farmhand I recognized but had never had a full conversation with. He rumbled up to the entrance of the shortcut, raised the tree, and then, with a *boom*, dropped it right across the path.

I blinked.

"Hey, uh, I think people are trying to get through there."

"Yeah," the farmhand growled. "That's the point, asshole. Nobody asked to use this field."

That's when I noticed the tracks. The rain had turned the grass into a churned-up mess, and suddenly, yeah, I could see why he was pissed. Not that it made the moment any less uncomfortable.

"I'm sorry," I offered. "I think some wires got crossed. People are just trying to move a family out today before their house gets condemned. We thought we had permission to go through"

"Oh, I know what's going on. And I'm so goddamn sick of you people," he spat. "I wish the flood took half the people in this town."

I tried to explain, fumbled through an apology, trying to express that it wasn't intentional. He told me, in no uncertain terms, to go sodomize myself.

Point taken.

We adjusted and found a new way. Everything was moved out of the house to where the family would be staying for the next few weeks. I still felt awful. Maybe the pasture fiasco wasn't directly my fault, but I'd definitely been captain of the good-intentions-bad-outcomes committee that day.

With my hat in hand and checkbook in my back pocket, I went to see the landowners to see if I could make things right.

The older couple, who lived there and owned the farm, were annoyed but ultimately understanding. When the topic of making good on the damaged hay came up, they waved me off and told me to speak with their son and daughter-in-law, who ran the farm operations these days. They were doing the evening milking out in the barn.

So I went out to the barn to finish smoothing over the situation.

Big mistake.

I didn't even get two steps into the milking parlor before the daughter-in-law spotted me and let loose. "Get the fuck out."

"Hey there, look, I just—"

"Get. The fuck. Out."

"I came here to make it right."

"You're not making anything right, asshole. Get the fuck out."

There's a certain kind of anger you can't argue with.

On the short ATV ride home, I felt like crap. What hit me hardest wasn't just the argument. It was the realization of how precarious things had already been for my neighbors, even before the flood. When I first moved to Peacham, there were seven dairy farms. By the time the waters rose, only three were left, and all of them were barely scraping by. Vermont's dairy farms once represented everything nostalgic and noble about agriculture: red barns, green hills, and multigenerational pride. But after World War II, the entire system shifted. Industrial-scale farms in the Midwest got bigger, more efficient, and heavily subsidized. Small New England dairies like these were left to die slowly.

And now, after the standoff at the barn, I could finally see what they were really angry about. It wasn't just the damaged pasture. It was everything that pasture symbolized: money they couldn't afford to lose, hard-won ground turned back into mud. And maybe worst of all, it was the realization that while volunteers and neighbors showed up to help flood victims in town, nobody was showing up for them and their struggling farm.

I couldn't shake the image of the daughter-in-law's face. The fury was obvious, but underneath it, something worse. Exhaustion, deep and bitter, the kind that doesn't come from one bad day but from years of watching things slip through your fingers. The kind that builds up over declining milk checks, climbing debt, and neighbors giving up, one by one. And then we came barreling in, all urgency and good intentions, trampling through their reality like we had a right to be there.

When I pulled into my driveway, the farm felt too quiet after the madness of the last two days. Alfred's excavator sat still, the dried mud on its tracks flaking off like burnt brownie edges. Toby Dog and Abby Dog gave me barks of greeting, blissfully unaware or maybe just indifferent to the fact that less than a mile away, someone's entire sense of control had been shattered.

I stepped out of the side-by-side, my hands still caked in dried mud, and just stood there. The day had started with a community racing to salvage what it could and ended with another family—my neighbors—feeling more isolated than ever.

It's easy to rally around a crisis. People show up. They help. They feel good about it. But what happens when the waters recede? When the adrenaline fades and the wreckage is less visible? That's when people really need help.

As I hope I have made abundantly clear this deep into this book, I am a person who has a severe lack of skills. No time made my inadequacy more apparent than those days after the flooding. I was out of my depth.

I couldn't rebuild homes. I couldn't unbury roads. I couldn't bring Dylan back.

But I could tell a story. And maybe, I could get people to listen and inspire them to help.

The idea started as a whisper in my head, half-baked and desperate: *Do a fundraiser. A livestream. Like last time, but bigger.*

I'd done a twenty-four-hour charity livestream before, raised roughly $25,000 for farmers in need when Vermont flooded in 2023. But that was different. That was for other people. This time, it was my town. My friends. My neighbors.

I started the livestream on a Saturday afternoon, sitting in my barn with a thermos of coffee and no plan except to keep going until I hit the twenty-four-hour mark. The first few hours felt slow, almost awkward. I told people about the flood. I showed them what was left behind. The waterlogged homes, the wrecked roads, the shell-shocked faces of my neighbors. I tried to explain what it felt like to stand in the mud and realize someone you knew wasn't coming back.

At first, the donations trickled in. Twenty bucks here. Fifty there. Kind messages in the chat. Encouraging words. I was grateful, but it still felt small. And then, things *changed*.

A few YouTubers I knew showed up. They spread the word. A couple of government officials called in. Then, Luis Guzmán (yes, *that* Luis Guzmán) joined in, talking about the years he spent living in our area.

By hour twelve, the chat was a frenzy. The donations weren't trickling anymore. They were pouring in, fast and constant, like people had just been waiting for a reason to give.

I didn't really sleep. I barely ate. The adrenaline carried me straight through to the next afternoon, and when I finally hit "End Stream," the final tally stared back at me like some impossible fever dream:

$120,000.

I sat there, staring at the screen, feeling like I had just come through another flood made of numbers and names and an overwhelming, impossible tide of generosity.

And that's when I saw it.

It was them.

My online community, the one I had spent *seven years* nurturing, had shown up to help the folks in my community. I had spent so long thinking of these two worlds as separate. Online vs. IRL. The town I lived in and the people I spoke to through a screen. But they weren't.

They had been connected this whole time. And I had never really seen it until now.

The money would help. That much was obvious. It would rebuild homes, replace furnaces, patch washed-out driveways, and keep farms from going under.

But the money wasn't the story. Not really. The real story was what it revealed. That community isn't just the people who share a ZIP code. It's not just the guys who show up with chainsaws after a storm or the neighbors who pull you out of a snowbank in January. It's bigger than that. It's the people who have never set foot in town but still felt the ground shift when the floodwaters came. It's the ones who had never met Dylan but still grieved for him, who had never enjoyed the tranquility of a summer day in Peacham but still wanted to put it back together.

And that's the irony, isn't it? Because when I first moved out here, I was chasing solitude. I wanted fewer people, less noise, more space. I was tired of cramming onto packed trains, of apartment walls so thin I knew my neighbors' breakup schedule better than my own. I wanted quiet. Escape. But here's the thing no one tells you: Moving to the country doesn't mean fewer relationships. It just means the ones you have matter more. Out here, when someone needs help, you don't just Venmo them twenty bucks and move on. You show up. You bring your tools. You wade into the mud. Because when you only have so many people, every single one of them matters.

For years, I kept my online world and my real world separate. The farm, the town, the people here—they were one thing. The folks watching through their screens were another. But I see it now. Community isn't about proximity. It's about who refuses to look away. Who carries the weight with you when you can't carry it alone. The flood took Dylan. And nothing makes that OK. But in its wake, it left something undeniable: proof that no one, not in this town, not in this life, has to go it alone.

CHAPTER 17

THE FARM WON'T SAVE YOU

The day I quit my final corporate job came in January of 2022. It should have been one of the happiest days of my life.

Since moving to Vermont back in 2018, I had had a day job. I ran the marketing department for a mid-size insurance company in Montpelier. It had been a significant pay cut from my DC job, but I had taken it because it meant I could move to the farm. In the background, I learned how to farm and start a business. Countless hours of burning candles at both ends. But no more! This was supposed to be the turning point. The moment when I left all of that behind.

No more corporate nonsense. No more bosses who micromanaged font choices on PowerPoint slides. No more pretending to care about synergy.

Just me. My farm. And my camera.

So why did I feel like I had just made a terrible mistake?

I shook off the thought, climbed into my truck, and pulled away, telling myself the unease was just nerves. Change was scary, after all. I had been hurtling down the same track for years, climbing, clawing, and chasing a version of success I never wanted. Of course, stepping off that track would feel disorienting. That was normal.

Despite the January chill, I rolled down the window, letting the wind slap against my face. I was heading home. Not to a cramped apartment in a city where I barely saw the sky, but to my farm. My farm. I had spent years building the dream in the margins of my corporate life. This wasn't just about escaping the job. It was about transformation. The farm would make me a different person. A better person.

The farm had fixed everything. That thought settled in my chest like a warm ember.

Then, ten minutes later, I pulled into a McDonald's drive-through.

I wasn't even hungry.

I ordered anyway. Two cheeseburgers, a large fry, orange Hi-C, ten-piece McNugget. A practiced and familiar order from nights spent drowning stress in salt and grease. The crumpled bag sat in the passenger seat as I drove home, the smell of hot oil and processed meat curling around me. I ate in silence, the rhythm of my chewing and swallowing the only sound in the truck cab.

Each bite was automatic. Mindless. The same way it had always been after a long, soul-draining workday. But I had quit my job. That life was over.

So why was I still doing this? To celebrate. I guess. Clean farm living from this day forward. Starting tomorrow.

Allison called as I drove home. Her voice was bright and supportive. "You're doing it," she said simply, warmly.

"We're doing it," I corrected gently, reminding myself it was our leap, not just mine. But after I hung up, a quiet, unwelcome thought surfaced: *What if the farm doesn't change anything?*

I swallowed hard, pushing it down with another bite of fries.

That maybe the problem wasn't my job.

Maybe the problem was me.

"OK, I guess I'll take…Morgan."

It was late fall, the air crisp and sharp, the kind that made your ears burn and your breath cloud in front of you. My sneakers were damp from the muddy patches near the makeshift goalposts made by our discarded backpacks. It was whoever was in our suburban Connecticut neighborhood who could get to the old baseball field. Most days, that meant two-hand touch. There were no standing teams. No coaches. No adults. Just a bunch of kids running plays we half-remembered from watching the NFL on Sundays. This was one of the only times I really felt like I belonged.

"Hey, Morgan, go long!"

I took off down the field, pumping my arms, feet pounding against the hard ground. The ball was already in the air. High, spiraling, perfect.

I reached for it.

And then, just as my fingers grazed the laces, something caught me from the side.

A body. A hit. I went down hard, my knee slamming into the dirt and my shoulder twisting underneath me.

Laughter.

I rolled onto my back, gasping, looking up at the cloudy sky. A shadow loomed over me.

"Damn," remarked Keith Offenberger, hands on his knees, grinning. "Felt like hitting a water balloon."

I forced a laugh and pushed myself up from the ground.

"Hey, you think he's got stuffing in there?" someone called from the sideline. "Like a beanbag?"

More laughter rippled through the group. Sharp, unforgiving, echoing in my ears long after they'd stopped. Each laugh felt like confirmation, an unspoken verdict about my place among them.

My face burned. I kept my head down, pretending to fix my shoelace.

"Or like one of those marshmallow Peeps," Robby added, nudging someone with his elbow. "Like, if you squish him, does he just puff right back up?"

It was a joke. Just a joke.

But I knew what it meant even if I wouldn't admit it.

I knew what they saw when they looked at me. I wasn't just another kid playing football anymore. I was *the fat kid playing football.* And there was no coming back from that.

I could hear them still laughing behind me, calling out as I left.

"Yo, Morgan, where you going? You quitting?!"

But I didn't turn around. I didn't stop until I had walked home. And then, I did the one thing I could think to do. I sat in the living room eating the leftover lasagna my mom had made the night before. It was supposed to be dinner for my sister and me that night, but it would have to suffice as an after-school snack.

And I ate. Not because I was hungry. Not because I wanted it. But because eating made the feeling stop. Food didn't whisper. Food didn't choose teams and pick you last. Food didn't laugh. Food didn't make fun of you.

I ate until the lasagna pan was scraped bare. I stared numbly at the smears of tomato sauce left behind, my stomach aching from fullness and emptiness at once. I sat there until the house grew quiet and dark.

As a result of this regular cycle, I gained more weight. And when you gain more weight, you get more ridicule. And when you get more ridicule, you feel more rejection. And by the time I graduated high school, it wasn't even a question anymore. I was *the fat guy*. Not a kid who happened to be overweight. Not someone who needed to "fill out" or "hit a growth spurt." Just fat. End of sentence.

I tried to change it. College was my first real attempt. I dropped a bunch of weight and got into lifting weights. I started to structure my entire life around food and fitness like I was a damn bodybuilder training for a cover shoot. I wasn't *working on* myself. I was *monitoring* myself. *Controlling* myself. Treating my body like a wayward employee who needed constant supervision and oversight. Because I wanted girls to like me and I didn't want to be a loser. I didn't want to be the guy people automatically assumed was the funny friend or the sidekick or the cautionary tale.

And for a while, I did okay. I had ups and downs, sure, but I was holding steady. Then my thirties happened. Career happened. And suddenly, my body was no longer my main project. It was just a thing I carried around while I focused on other things. I still tried to work out, but good luck maintaining a strict gym schedule when living out of suitcases, sitting through business dinners, and stress-eating drive-through burgers at 11:45 p.m. while drowning in meetings and last-minute marketing plans.

At first, I told myself it was just life. Work, stress, exhaustion. The usual things that make people gain weight. I was just looking for something else to blame.

I started getting into organic food, farmers' markets, and regenerative agriculture. My plan was that if I could just get rid of *the bad food* and swap out the neon-lit fast food for homegrown produce, then that could fix everything. Perhaps I wasn't out of control. Maybe I was just *poisoned* by the modern world.

If only I could escape it all. Get back to *real* food, *real* land, *real* work. Maybe that would be the thing that could save me. Maybe that would break the cycle. Or maybe it wouldn't. But it was easier to believe in some kind of food-based salvation than to admit that my problems were never just about the food.

Tuesday afternoon. Early mud season. That weird, in-between season where winter is technically over but the cold still clings to the ground like an ex who won't move on. The air smelled like damp earth and thawing manure, a combo that somehow made my eyes water *and* my stomach growl.

I had a dozen things on my list that day. Farm chores now became the priority since I was no longer working a day job. Feed the cows, fix the fence, and scoop a truly horrifying amount of chicken poops. But first, I had to deal with *her*.

Belinda Carlisle. The escape artist.

One of last year's calves, she had slipped through a gap in the fence *again*. I spotted her hoof prints in the mud, little crescent-shaped insults leading straight into the woods. I swore under my breath, because I had checked the damn fence this morning. *Personally.*

Boots sinking into the muck, I followed the trail, and there she was, just beyond the tree line, chewing on a patch of winter grass like she had *nothing better to do*.

"Come on," I muttered. "Let's go."

She flicked an ear. Looked at me. Then bolted deeper into the trees.

I ran after her, nearly eating a faceful of mud in the process, breath coming harder than it should have. Not because I was out of shape (okay, *maybe* a little because I was out of shape), but because this was so goshdarn *typical.*

The farm was supposed to be different. Predictable. *Fixable.* But here I was, chasing a runaway heifer through the woods like an idiot, wondering how the hell *any* of this was supposed to fix *me*.

My hands were shaking when I finally cornered her near a fallen log. Not from exhaustion. Not from fear. From *anger*.

"Why," I panted, stepping closer, "can't you just stay where you're *supposed to*?"

She blinked at me. Wide-eyed. Unbothered. I exhaled sharply, rolling out my shoulders, trying to *unclench*. This was about control.

I had spent my whole life wrestling control over *something*. My habits, my weight, my career, my future. I had traded corporate spreadsheets for pasture rotations, desk jobs for farm chores. But I was still losing a battle of control.

Because the farm didn't care about my plans. The heifer didn't care about my expectations. And those feelings of shame, inadequacy, and the relentless hunger for approval I'd been carrying since childhood sure as hell didn't respect my illusions of control. The realization hit like electrified fence wire in the face: The farm wouldn't transform me into someone else. It was just another place to hide.

I could build the perfect farm. The perfect diet. The perfect version of myself. And I'd still be me, with all my quirks and flaws.

Still the kid walking off the football field, burning with shame. Still the guy sitting in his car outside a gas station, unwrapping a burger like it was contraband. Still convincing myself that *this time*—this job, this diet, this farm—would be the thing that finally fixed me.

The heifer snorted, snapping me back to reality. She shifted her weight, watching me.

Waiting.

I let out a slow breath. Then another.

"Alright," I muttered, stepping back. "I give up. Let's go home."

And miraculously, she started following me. Looking back on that moment, I like to think it was my sense of surrender and release of control that encouraged her to follow me. But more likely it was the bucket of alfalfa treats I was carrying.

When I had worked in an office, when my boss had called me in to tweak some meaningless detail on a project, when I had stayed late and answered emails from bed just to prove I was someone worth keeping—hadn't it been the same? The same tightening in my gut, the same feeling of slipping, of losing control?

I had left that world behind. But it hadn't left *me*. And at this point in my life, working full-time on the farm and making videos that made me feel creatively fulfilled, I had no more bosses or jobs or playground bullies to blame. It was just me. I was the only one left to take accountability.

And that's when, at the age of forty-two, I started seeing a therapist to help me work on my binge eating disorder.

Men don't talk about binge eating disorders, but not because they're rare. Eating disorders still usually get boxed in as a "women's issue," so when a guy loses control around food, it's not seen as a disorder.

"He's just a guy who loves to eat."

"The dude just lacks discipline and willpower."

"We all gotta deal with stress in our own ways."

But binge eating disorder isn't just eating too much. It's a compulsion. A cycle. A dopamine loop that hijacks the brain, turning food into a quick, predictable high the same way some people chase booze, gambling, or video games. The craving kicks in, the binge happens, and then comes the shame, the regret, and the promise to *do better next time.* Except the next time always comes. And because most men with binge eating disorder don't look like the stereotype of an eating disorder: gaunt, frail, wasting away. No one thinks to call it what it is. Instead, it's chalked up to a moral failing. Laziness. Lack of willpower. Bad habits. And so they don't seek help. They don't talk about it. They just marinate in the shame. They just keep fighting the same battle alone, trapped in a pattern that wrecks their health, confidence, and sense of control.

When I started to look at my problems from that perspective, many things clicked into place all at once. Because at the end of the day, that's what it was. *Compulsion.* Food was the easiest, most accessible dopamine hit I could get. It was reliable. It never required effort or motivation. It was always there, ready to drown out the thoughts, to fill the space when I couldn't.

I wasn't diagnosed with ADHD as a kid, even though my mom always had her suspicions. In my early twenties, the label was finally clinically applied. And sure, people love to roll their eyes at it. Call it over-diagnosed. Blame it on too many glowing rectangles. But I know what my brain does. Or, more accurately, what it *doesn't* do.

ADHD isn't just about struggling to pay attention. It's also about dopamine. About chasing that next hit. About latching onto whatever makes your brain light up and then drowning yourself in it. That's the fight.

I've made a few different videos over the years talking about how I also have a video game addiction. I've lost years of my life living as Niko Bellic or Arthur Morgan. I deal with this problem by not playing video games. And while I've had the occasional relapse, it's not too hard to avoid playing video games. I just have to commit to not playing video games and can stick to it.

But with food, it's not so simple. A person can live their entire life without playing video games and be just fine. If you try that same approach with food, it will kill you after a few weeks. You can't quit eating.

A binge doesn't start with hunger. It starts with an idea. A pizza. Onion rings. Pizza with onion rings! The thought sneaks in, quiet at first. Then it gets louder. *Louder.* Until suddenly, I'm not thinking about anything else. Not the work I need to do. Not the task in front of me. Just the noise. Just the planning. How to get it. Where to go. How to be *alone* when I eat it.

And yeah, I'm aware of it now. I see the pattern. I understand the trick my brain is playing. But that doesn't stop it. Knowing makes it worse because now I'm not just fighting the impulse. I'm fighting the shame of seeing it happen in real time and still being unable to stop it.

That night, I sat down at my computer and looked up a therapist.

Not because I had finally "hit rock bottom." Not because I needed some dramatic intervention. But because, for the first time, I wasn't trying to fix myself. I was just trying to understand. And maybe that was the point.

The next morning, I woke up and immediately regretted it. Not the therapy part. That was still scheduled for a few days away. No, I regretted letting myself admit, even for a second, that I needed help. That I couldn't just muscle through this like I had everything else.

I lay there, staring at the ceiling, bargaining with myself.

Maybe I was just tired. Maybe I just needed to get back into a routine. Maybe I didn't need therapy. Just *discipline.*

Discipline had always been my answer. It got me through weight loss in college, tracking every macro, forcing myself to the gym, ignoring hunger like it was background noise. It got me through my career, first one in, last one out, making myself indispensable. And it got me through the farm, early mornings, long days, everything stacked with efficiency in mind.

If I could just *lock in*, I wouldn't need some therapist telling me what I already knew.

I threw off the blankets. Today, I'd prove it.

I attacked the morning like I was training for the Olympics. Fed the animals. Cleaned the barn. Organized the tool shed for no reason.

Moved through chores with mechanical precision. Didn't stop for breakfast. Didn't stop for anything.

The thing about discipline? It *feels* like it's working, right up until the moment it isn't.

By midday, I was running on fumes. And then, *the thought.* That small, insidious voice. Do you know what would fix this? The answer was always food. Not a meal. Not something reasonable. Something big. Something *fast.* Something to drown out whatever was clawing at me from the inside.

I shoved my hands into my pockets, forcing myself to focus on the fence I was supposed to be fixing. But the cycle had already started. I tried to push the thought away. That only made it louder. Burger. Fries. Onion rings. A milkshake. No, *two.* You deserve it. You *need* it.

My brain was already pulling up the playbook. Mapping out where to go, how long it would take, how much food I could order without it looking suspicious. This wasn't hunger. This was compulsion. The dopamine loop. The same script I'd been following since I was a kid: eat fast, eat alone, eat like I could outrun whatever I was feeling.

And even now, *knowing* what was happening in my brain, I still couldn't turn it off. Which meant one thing: Discipline wasn't going to save me.

I gripped the fence post and took a deep breath.

Two options.

1. Keep doing what I'd always done. White-knuckle through it, pretend I had control, then inevitably cave hours later when my willpower gave out.
2. Try something different.

I didn't know what *different* looked like. But I knew what *the same* looked like. So instead of getting in my truck, instead of running the same tired play, I did something else. I sat down at the dining room table and opened a notebook. I began writing. Not organized. Not neat. Not narrative. Just raw and honest scribbling. I wrote about the shame, the cravings, the constant whisper of inadequacy that had followed me from childhood. With each sentence, the urgency of the binge dulled slightly, replaced by something unfamiliar: clarity. For once, I let myself feel instead of burying those feelings beneath food.

The thoughts didn't stop. The craving didn't vanish. But I didn't *feed* it.

I let it sit. Let it burn. Let my body feel it without trying to smother it with food. And word by word, it lost some of its power. Not all of it. Not even most of it. But just enough. Enough to remind me that I could sit with discomfort and survive it. That I didn't *have* to numb it. That I could *want* something and not give in to it.

By the time I got back outside, nothing had changed.

The sun was still shining. The animals were still there. The work still needed doing.

But *I* had changed. Not a lot. Not enough to make a neat little narrative out of. But enough. I stayed in it. The discomfort, the temptation to check out, the pull toward grease and salt. And nothing happened. No epiphany. No cinematic breakthrough. I just stayed. Which meant, for the first time, I didn't have to fight myself for dominance. Coexisting was an option. I could be the same mess, just with slightly less resistance.

That would have to be enough.

CONCLUSION
LIVING DIRT RICH

She was dead.

It was a lazy summer afternoon, and I had just found Jemima Puddleduck lying under an apple tree that I had planted myself. Her white feathers lifted slightly in the breeze, catching the sunlight, but she was still. Her eyes, usually so tender, were cold and glassy.

Jemima Puddleduck was the farm's matriarch. The perpetual grandmother of our flocks. A widow, her husband, Samuel Puddleduck, had grown so massive he could barely walk. I ended up having to butcher him for the dinner table before he started to suffer. Their son, Samson, followed in his father's overgrown footsteps, becoming the largest duck I'd ever raised. He ended up meeting the same fate as his father. Jemima's daughter, Delilah Puddleduck, was taken by a bobcat shortly before Toby Dog arrived on the farm to watch over the flock.

Jemima had been slowing down all summer. Keeping to herself. Losing weight. I'd known this was coming. I'd told myself I was ready for it. But death, even when expected, has a way of sneaking up and punching you in the stomach. Particularly when you've had a duck for more than six years. It was her eggs that were the ones to teach you that you were, in fact, allergic to duck eggs.

Standing over Jemima's lifeless body, I felt the weight of all those stories. She had lived through so much of our farm's story. The small victories, the bad decisions, the moments I wish I could redo. And now, in the stillness of the orchard, she marked another shift. A quiet closing of another chapter.

By now, I've buried more ducks than I can count. The first few left me wrecked, stomach clenched, wondering if I was built for this. Loss now feels different. Not easier. Just familiar. Like a returning season. It doesn't stop me in my tracks the way it used to. But it also doesn't hurt less.

I put her body where I put most of my dead poultry: On the far back corner of the farm, as far as possible from the bird yard, just so the coyotes or vultures or whoever would scavenge her corpse wouldn't go getting any ideas. Afterward, I sat momentarily, watching the sky shift to a deeper blue, thinking about how farm life never moves in straight lines. It loops and doubles back, pushing through growth, loss, mistakes, small triumphs, and another round of the same. The wheel keeps turning. Again and again.

The world loves a tidy ending. Our story-driven minds crave a clean finish. The hero victorious, the lost found, the puzzle solved. But farms don't deal in neat resolutions. They deal in cycles. Seasons, storms, births, burials. Nothing here ever really ends. It just gets out of the way for whatever is coming next.

That's what *dirt rich living* really is. Not simple. Not peaceful. But honest. It's letting the dirt get under your fingernails and still sitting down to dinner at the end of the day with a full and open heart. It's the kind of richness that doesn't show up in your bank account but in how much of yourself you're willing to give to something that doesn't promise anything back.

There is a toxic misconception that farm life is simple, an antidote to modern chaos. A pastoral escape from the grind of screens, deadlines, and too many other people. You picture yourself relaxing on the porch, coffee in hand, watching the sunrise spill gold across your land while chickens scratch in the dirt. And yeah, sometimes, it's exactly that.

But mostly, it's not.

Farms don't care if you're exhausted, bleeding, or halfway to a breakdown. The animals still need to eat. The fences won't fix themselves. The weather will do whatever the hell it wants. One season tricks you into thinking you've cracked the code. The next reminds you that you're just another creature in the dirt, trying to survive.

If you come to farming looking for meaning, you'll find it. But you won't find it in grand epiphanies or Instagram-worthy moments. You'll find it in the mess. You'll find it in the dirt. Out of the dirt we came, and eventually, into the dirt we go.

If you dream of farming, understand this: The joy and the heartbreak come as a package deal. You will stand in quiet awe. The first ripe tomato of the season, a calf wobbling to its feet, the low hum of crickets in a warm summer field. And you will stand in gut-punched grief—a predator wiping out your flock, a duckling that fails to thrive, a windstorm that rips apart your greenhouse. Nobody tells you how much loss is built into the life. But loss is part of the contract.

So how do you prepare? First, drop the idea that you're ever in control. Nature has the final say. Your job isn't to conquer it. It's to listen. You should always do your best. And sometimes that won't be enough.

And so here we are at the "end" of a story that doesn't really end. A story of leaving the corporate world, wrestling with old habits, forging a life from the soil, and learning over and over that contentment isn't something you build out of circumstances. It's something you find inside yourself when you finally stop trying to bend the world to your will.

Tomorrow, I'll be up before dawn. I'll feed the animals, pour the water, and scratch Toby Dog and Abby Dog behind the ears. I'll probably glance at the barn, half-expecting to see Molly Murder Mittens perched on a fence post, eyes flashing a warm greeting. And my heart will ache at the sight of a cat-less fence post. I'll fill the feeders for the ducks, wave to a passing neighbor, and maybe record a few clips for the next story I share. And somewhere in between, I'll think about friends dead and gone, and wish that they could see the sunrise too.

And, of course, something will go wrong. A hose will freeze. A bird will be sick. I'll do something stupid. I'll wrestle with the old instincts: the doubt, the urge to *fix it all, right now*. But by evening, I'll have lived another day on this land. Imperfectly and wholeheartedly.

And that's the thing no glossy magazine, dreamy Instagram story, or lame influencer-authored book is going to tell you:

You don't ever arrive at your destination. You just keep turning the wheel. Over and over and over. And that is more than enough.

ACKNOWLEDGMENTS

Writing a book is a lot like farming: messy, uncertain, and occasionally painful. And it would be impossible without plenty of help.

First and foremost, thank you to Allison. You didn't just humor my crazy dream. You fully embraced it. Your patience, humor, intelligence, and occasional eye rolls have kept me grounded and inspired. I'm beyond lucky to share this life with a weirdo like you.

To my family and friends, thanks for always believing in me, even when you weren't entirely sure what the heck I was up to. Your unwavering encouragement and gentle reminders about what matters most shaped not only this book, but the person who wrote it.

An enormous thank you to my dear friend Jessica Sowards, whose feedback, honesty, and enthusiasm made her my constant reader and invaluable sounding board throughout the writing process. Your heart is true. You're a pal and a confidant.

To my literary agents…Alan Nevins and Jacklyn Saferstein-Hansen… your guidance, faith, and support have been critical in bringing this book to life. Thanks for believing in my voice and helping to share this story.

To Charlie Button and the entire Select team…Lauren Rosenzweig, Sam Rivman, Matt Teng, Nikki Schwartz, and Sharvari Bhat…thank you for helping me navigate the twists and turns of this strange career. Your wisdom and support allow me to focus on what I love most: telling stories and taking care of the land and the animals.

Speaking of animals…special thanks to Toby Dog, Abby Dog, Pablo Barn Cat, Lil Barn Cat, Ginny Barn Cat, and the dearly departed Molly Murder Mittens. You're the best gang of co-workers a guy could have. To the ducks, geese, cattle, gohtwas, pigs, chickens, bees, trees, and everything else wandering, waddling, mooing, and buzzing through our lives, thank you for your patience, comic relief, occasional chaos, and constant lessons in humility. Life without you would be quieter, cleaner, and infinitely more boring.

To the Peacham community, and all my friends and neighbors in Vermont…thank you for your warmth, generosity, and endless patience with the weird newcomer who showed up asking way too many questions. You've made this place truly feel like home, and I'm grateful to be here every single day.

Finally, to every single person who's followed along with our little farm online…wherever and however you found your way here…thank you. Your weirdness, empathy, curiosity, advice, and encouragement have meant more than you'll ever know. Without you watching, sharing, supporting, and genuinely caring, this book wouldn't exist. Your support has transformed a farm fever dream into something meaningful, real, and worth sharing. I'm forever grateful.

From one mess to another,

Morgan Gold
Peacham, VT